GIS and Organizations

HEATHER CAMPBELL AND IAN MASSER

Department of Town and Regional Planning, University of Sheffield, Sheffield, UK

| UK | Taylor & Francis Ltd, 4 John Street, London WC1N 2ET |
| USA | Taylor & Francis Inc., 1900 Frost Road, Suite 101, Bristol, PA 19007 |

Copyright © Heather Campbell and Ian Masser 1995

All rights reserved. No part of this publication may be reproduced, stored in a retrieval system, or transmitted, in any form or by any means, electronic, electrostatic, magnetic tape, mechanical, photocopying, recording or otherwise, without the prior permission of the copyright owner.

British Library Cataloguing in Publication Data

A catalogue record for this book is available from the British Library

ISBN 0 7484 02047 (cased)
ISBN 0 7484 02055 (paper)

Library of Congress Cataloging in Publication Data are available

Cover design by Hybert Design and Type, Maidenhead, UK

Typeset by Solidus, Bristol, UK

Printed in Great Britain by Burgess Science Press, Basingstoke, on paper which has a specified pH value on final paper manufacture of not less than 7.5 and is therefore 'acid free'.

11765267

Contents

List of figures and tables		vii
Acknowledgements		ix
1	**Introduction**	**1**
	Technology: innovation or irrelevance?	1
	Diffusion	4
	Overview	8
2	**Technology and organizations**	**9**
	Introduction	9
	Technology	9
	Organizations	13
	Reinvention	23
	Conclusion	24
3	**Perspectives on implementation**	**25**
	Introduction	25
	Technological determinism	26
	Managerial rationalism	29
	Social interactionism	34
	Success and failure	49
	Conclusion	50
4	**Organizations and GIS: a case study of British local government**	**51**
	Introduction	51
	British local government	52

Contents

	Research strategy	61
	Overview of the research findings	64
5	**The diffusion of GIS in British local government**	**65**
	Introduction	65
	GIS adoption	65
	System development	68
	Choice of technology	72
	Benefits and problems associated with GIS	74
	Main changes since 1991	76
	Evaluation	79
	Conclusion	83
6	**Reinvention and utilization: GIS in practice**	**84**
	Introduction	84
	The case studies	85
	Why do organizations adopt GIS?	87
	Characteristics of the GIS technologies being implemented in the case study authorities	90
	Evaluation	108
7	**GIS implementation: a problematic process**	**113**
	Introduction	113
	Technological considerations	114
	Organizational considerations	118
	Information management strategies	118
	Commitment and participation	132
	Instability	144
	Conclusion	151
8	**GIS: innovation or irrelevance?**	**153**
	Introduction	153
	Implications of the research findings	154
	Implications for research	161
	Conclusion	162
	References	164
	Index	173

Figures and tables

Figures

1.1	A conceptualization of diffusion	6
3.1	Underlying rationale of the corporate approach to GIS implementation	33
3.2	Limitations in the underlying rationale of the corporate approach to GIS implementation	42
4.1	District authorities in Great Britain in 1994	53
4.2	County and regional authorities in Great Britain in 1994	54

Tables

2.1	Key characteristics of a selection of theoretical approaches to decision-making in organizations	20
3.1	Main characteristics of technological determinism	26
3.2	Main characteristics of managerial rationalism	29
3.3	Arguments in favour of a corporate strategy towards GIS implementation	34
3.4	Main characteristics of social interactionism	37
3.5	Arguments that question the appropriateness of a corporate strategy towards GIS implementation	45
4.1	Average population of local authorities in Great Britain in 1991	55
5.1	Plans for GIS in local authorities in Great Britain	66
5.2	Authorities with GIS by type of local authority	66
5.3	Percentage of local authorities with GIS facilities by type and region	67

Figures and tables

5.4	Level of GIS adoption by authorities and systems	68
5.5	Length of experience with GIS technologies	69
5.6	Approach to GIS implementation	69
5.7	Number of departments involved in multi-departmental GIS facilities	70
5.8	Departments involved in GIS facilities	70
5.9	Lead departments in multi-departmental facilities	71
5.10	Single department GIS	72
5.11	Software adopted for GIS work	73
5.12	Hardware adopted for GIS work	74
5.13	Most important benefits associated with GIS	75
5.14	Most important problems associated with GIS	75
5.15	Summary of main changes between 1991 and 1993 with respect to GIS adoption and system development	76
5.16	Summary of main changes between 1991 and 1993 with respect to GIS technology and perceived benefits and problems	78
6.1	Characteristics of the case studies	86
6.2	Technical characteristics of the GIS adopted by the case study authorities	91
6.3	Number of terminals able to access GIS technology	93
6.4	Types of departments with computer access to GIS facilities	94
6.5	Range of skills and number of staff employed in implementing GIS technology	96
6.6	Organizational structure for GIS implementation	99
6.7	Number of departments involved and with access to GIS facilities	100
6.8	Departments with attribute data on a GIS	103
6.9	Operational applications	105
6.10	Applications under development	105
6.11	Overview of the utilization of GIS in the case studies	107
7.1	Perspectives on the implementation process	114

Acknowledgements

This book, like the implementation of a geographic information system, has been dependent upon the efforts and inspiration of a great many unsung individuals and supportive organizations. A huge debt is owed to the very many people in local authorities across the length and breadth of Great Britain who gave unsparingly of their time and insight. Particular thanks must be extended to those authorities who consented to act as case studies. Their willingness to facilitate the research and to provide open access to documentary records and precious staff time was invaluable. The ideas developed in this book stem from their experiences, although responsibility for the interpretation is our own and should not be attributed to any particular individual or local authority.

The patience and good humour of our research assistants who had the unenviable task of conducting two complete telephone surveys of British local authorities undoubtedly made our task easier. The skill and dedication shown by Lee Garnett, Jenny Poxon, Elizabeth Sharp and Max Craglia made a significant contribution to this project. In this connection we must also acknowledge the financial support provided through the Economic and Social Research Council/ Natural Environment Research Council's joint initiative on Geographic Information Handling and from the Local Government Management Board, without which the research providing the focus of this book would not have been possible.

The final compilation of this book, however, is the result of the tolerance of colleagues in the Department of Town and Regional Planning at Sheffield. In particular we are very appreciative of the efforts of Dale Shaw, Melanie Holdsworth and Christine Goacher for unscrambling our text and Christine Openshaw and Graham Allsopp in the preparation of some of the tables and figures. The support of our publisher Richard Steele and his team at Taylor & Francis has also greatly assisted the production of this book.

Acknowledgements

Finally and most importantly the encouragement and forbearance of family and friends plays an incalculable part in a project of this nature. We are indebted to you.

CHAPTER 1

Introduction

Technology: innovation or irrelevance?

The ability to innovate is generally regarded as fundamental to organizational survival. Governments throughout the world are expending considerable resources searching for technological innovations and novel techniques which it is assumed will increase industrial as well as administrative competitiveness. However, many seemingly good ideas remain in the workshops of their inventors. Some innovations diffuse rapidly throughout society while other equally worthy developments progress little further than the laboratory. These observations question the extent to which diffusion is dependent upon the inherent technological characteristics of a particular innovation. At least as important appear to be the cultural, organizational and institutional contexts into which such a development is to be embedded (Bijker, Hughes and Pinch 1987; Dunlop and Kling 1991a; Goodman, Sproull and Associates 1990; Rogers 1983). It is, therefore, crucial that consideration is not only given to the elegance of the technology but the interrelationships between organizations and innovations if greater understanding is to be achieved of how a potentially good idea becomes a facility taken for granted. As a result, this study seeks to investigate how organizations respond to innovations and, more particularly, what processes influence the effective implementation of new technologies in real-world situations.

The innovation that provides the focus for the study is geographic information systems (GIS) and the organizational context is British local government. GIS are essentially a set of computer-based technologies which are able to store, display, manipulate and analyse spatial data, most particularly map-based information. Arguably, manual forms of GIS have been available for many decades and computer-based systems since the 1960s. However, recent advances in the data-handling capabilities of computers, most particularly the speed with which large data sets can be processed, led GIS to become commercially viable during the 1980s. As a result, the concepts on which GIS are based are by no means new, but the speed and flexibility with which it is now possible to exploit the geographical properties of information has produced an innovation that has prompted considerable interest as well as prophetic claims for the technology. It is not only vendors, manufacturers and researchers who have been enthusiastic about GIS technologies; developments in this field led the British government to commission an inquiry chaired by Lord Chorley to investigate the issues surrounding the handling of geographic information. The landmark report of this Commission was published in 1987 with the significance of the technological advances they had seen encapsulated in the following now legendary statement: 'Such a system is as significant to spatial analysis as the inventions of the microscope and telescope were to science, the computer to economics and the printing press to information dissemination. It is the biggest step forward in the handling of geographic information since the invention of the map' (Department of the Environment 1987, para.1.7).

The conceptually simple capacity to be able to unleash the power of computers to process spatial data and thereby to use geography as the main organizing principle for database design has an appeal which, in conjunction with the map-based outputs, started to capture the imagination in the 1980s. GIS technologies were promoted as offering opportunities to improve efficiency by reducing the duplication of spatial data sets and at the same time ensuring that all sections of an organization had access to the same up-to-date information. Furthermore, GIS were envisaged as contributing to organizational effectiveness through the provision of basic data as well as stimulating more complex analyses which would enhance decision-making capabilities at operational, managerial and strategic levels. Fundamental to such claims was the capacity of GIS to integrate data sets from a wide variety of sources. As a result, GIS were regarded as facilitating data sharing which in turn would lead to more informed decision-making and logically better decisions. Some went further, suggesting that the greater availability of information would result in a democratization of decision-making and improvements to the quality of life of the whole of society.

The optimism associated with the potential of GIS mirrors much that has been written in relation to information technology in general. Underlying such sentiments is a strong feeling that improvements in social and economic conditions are dependent upon technological innovations, most particularly developments in computing (Feigenbaum and McCorduck 1991; Naisbitt 1984; Toffler 1980). Implicit within the utopian views of technology is an assumption that society is entering, or perhaps has already entered, an information age, in which economic prosperity as well as harmony within society depends upon the ability to process vast reserves of information. Toffler (1980) terms this the 'Third Wave', a period in which successful market economies are distinguished by the rapid diffusion of computers. As a result, it is predicted that mechanical production methods will gradually be replaced by programmable technologies.

Introduction

The increasing availability of information in association with the ability to exploit these resources to the full is not only envisaged as a vital contributor to economic success but also as liberating those who have traditionally felt excluded from the decisions that affect their lives. For example, in the workplace computerized technology has been conceptualized as 'informating' activities, thereby removing the tedium from many tasks and dispersing power more widely (Zuboff 1988).

Despite claims about the potential opportunities that will result from the introduction of new forms of computer-based systems such as GIS, very little is known about the actual impact that this technology is having in practice. As yet, attention has largely concentrated on enhancing technical know-how in relation to handling geographic information. Technological innovation is, however, of little value if the product of such developments proves problematic to implement or is regarded as an irrelevance by potential users. In contrast to the expectations of the utopian conceptions of technology, evidence suggests that technical elegance is no guarantor of widespread utilization. For example, studies investigating the introduction of computer-based systems, mainly within the private sector, have shown that marginal gains, unforeseen problems or even complete abandonment of the project are more common than success (Dunlop and Kling 1991c; Eason 1988; Lucas 1975; Lyytinen and Hirschheim 1987; Moore 1993; Mowshowitz 1976). The objective of this book, therefore, is to examine the relationship between an innovation and the organizational context in which it is to be embedded. Is the prime determinant of the widespread diffusion of a new technology its technical capabilities or are other factors responsible for the outcome of this process? Is the dawning of the information age, if that is what is taking place, about technological innovation or the capacity of organizations to absorb change?

GIS technologies provide an excellent case study for such an investigation. They exemplify the characteristics of the new generation of programmable technologies which some suggest will transform society. The potential of this technology has been eloquently outlined but how are users reacting in practice? It is this group, whose concerns have little to do with GIS and are therefore agnostic and perhaps even sceptical about the merits of changes to existing work practices, that will make the ultimate judgement as to its value. Will their verdict be innovation or irrelevance, and what processes will contribute to this assessment?

The current stage of development of GIS technologies makes this an opportune time to examine the experiences of users. The initial purchase of any new product is usually associated with a great deal of enthusiasm and expectation. It is therefore important before embarking on an investigation of this type that sufficient time has passed to allow a realistic assessment to be made. In the case of GIS such circumstances now appear to have been fulfilled as early adopters introduced the technology into their organizations in the mid-1980s. Such organizations are by their very nature often viewed as atypical. However, it is now possible to draw on the experiences of what may be termed the second generation of GIS adopters, without falling into the trap of being forced to focus on the very early stages of implementation in these organizations.

The organizational environment that provides the backcloth for the introduction of GIS, is British local government. If computer-based technologies are about facilitating information handling, where better to examine the input of one group of these innovations than in a set of organizations whose *raison d'être* is focused on consuming, processing, managing and manufacturing information. As Hoggett rather neatly puts it, 'what poultry is to Kentucky Fried Chicken, information is to local government' (Hoggett 1987, p. 226). Furthermore, not only does information appear to be the focus of the majority of activities of local government but also, according to the Chorley Report, at least 60 per cent has a geographical component (Department of the Environment 1987). Local government, therefore, seems to be an appropriate context in which to examine the potential of GIS and more particularly the interrelationships between innovations and organizations.

The process underlying much of the preceeding discussion has been the concept of diffusion. The remainder of this chapter will consider the nature of diffusion in relation to a technological innovation such as GIS. This will be followed by an overview of the structure of the book.

Diffusion

Diffusion is the fundamental process that is responsible for the transfer of innovations from the workshops of their inventors to becoming a daily part of the lives of a large section of society. There appears to be a great deal of ambiguity surrounding the precise meaning of diffusion. It is possible to identify two groups of researchers for which the concept of diffusion has provided a focus to their studies. First, geographers have shown some interest in the spatial aspects of diffusion. Their principal concern has been to investigate the effect of distance from the source of an innovation on the speed and extent of adoption (see, for example, Hägerstrand 1952 and 1967). It is, however, arguable that it is the second group of researchers based in the sociology discipline who have made the most significant contribution to our understanding of the process of diffusion (see, for example, Eveland, Klepper and Rogers 1977; Rogers 1983 and 1986).

Rogers in his overview of research in this field defines diffusion as 'the process by which an innovation is communicated through certain channels over time among members of a social system. It is a special type of communication, in that the messages are concerned with new ideas' (Rogers 1983, p. 5). This definition is important as it identifies a number of significant concepts in relation to diffusion. The first concerns the innovative nature of the product being diffused and therefore the uncertainty surrounding the consequences of adoption. The second important element of the definition is the stress placed on the whole

Introduction

role of communication channels as the means of relaying information about the merits and deficiencies of an innovation. These channels include both the mass media and the opinions of peers, in particular what are termed 'opinion leaders'. Rogers also emphasizes the significance of time as well as the nature of the social system in which the process of diffusion is located. The concepts embedded within this overall conceptualization are important, as the speed and extent of the diffusion of an innovation is linked to social and political processes rather than simply the inherent technical worth of the product.

Practitioners in this field acknowledge, however, that there are a number of limitations to the scope of much of the existing empirical research in this field (Rogers 1993). For instance, implicit within much of this work is a pro-innovation bias, that is to say an underlying assumption that the adoption of the particular development under consideration will necessarily benefit all concerned. Perhaps, more importantly in relation to a technology such as GIS, there has been a tendency to concentrate on relatively simple forms of diffusion involving decisions based solely on individual adopters. Ryan and Gross's (1943) landmark study, for instance, concentrates on the adoption of hybrid seed corn by farmers in two Iowa communities. Even in relation to the recent developments in computing and telecommunications capabilities, diffusion studies have tended to concentrate on products designed to meet the needs of the personal end of this market. For example, Dutton, Rogers and Suk-ho Jun (1987) have examined trends in home computing. Technologies such as GIS, however, present a more complicated situation, for rather than being adopted by one individual, the unit of analysis is usually an organization. Decisions in this type of context are therefore the result of the interaction between complex sets of personal, organizational and even cultural interests. Furthermore the purchase of the innovation is not the culmination of this process as there is no guarantee that the equipment that has been acquired will actually be utilized within the host organization. Consequently, it would appear that diffusion should be conceptualized as consisting of, first, a set of processes associated with the propensity of individuals or organizations to adopt a particular technology, and second, a set of processes that influence the effective utilization of an innovation. While the processes involved in both are likely to be closely related and mutually dependent, a useful conceptual device would be to view the former as essentially external to the organization while the latter is concerned with diffusion within the organization.

A further complication in devising an operational conceptualization of diffusion is that there is a tendency for it to be synonymous with terms such as adoption, implementation, routinization and utilization without any clear distinction as to the relationship between these concepts. Given that the innovation providing the focus for this study is GIS, it seems to be appropriate to regard diffusion as an umbrella concept: a term that encapsulates the processes of awareness raising, adoption, implementation, routinization and utilization and an assessment of the consequences of the entire exercise for the individuals and organizations concerned (see Figure 1.1). It is important, however, that while all

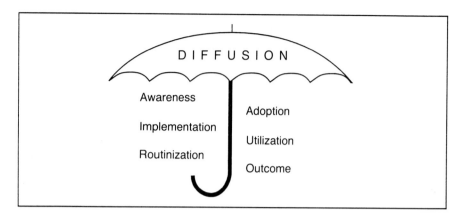

Figure 1.1 A conceptualization of diffusion

these elements are regarded as an inherent part of diffusion, the overall process is not considered to be linear in nature. It is a complex and problematic process in which aspects of implementation may well precede adoption: while one section of an organization may be routinely utilizing a technology, another may be unaware of its existence. Furthermore, in the case of a programmable technology such as GIS it is probable that there will be repeated cycles of development, learning and use as circumstances change or users become more demanding. Diffusion is not therefore a once-and-for-all process.

In contrast to much of the existing research on the diffusion of innovations, this study not only examines the speed and extent of GIS adoption within one organizational grouping but more particularly concentrates on the processes that influence whether effective utilization is achieved. Leonard-Barton (1988) notes that there has been a tendency for research to focus on the initial decision to adopt a new form of technology, ignoring the critical phase of implementation which underpins the transition from untested expectations to a taken-for-granted technology. This emphasis on exploring the relationship between organizations and innovations complements and supplements work on GIS which is currently under way, focusing on the traditional concerns of diffusion studies, most particularly the external communications channels through which knowledge about an innovation diffuses (see Assimakopoulos 1995). An increasing number of high-profile and often extremely expensive cases of information systems that have had to be abandoned point to the problematic nature of implementation (South West Thames Regional Health Authority 1993). The Audit Commission (1994) in a recent report has highlighted the poor return that many public organizations such as health and local authorities have made on their investments. It is also noticeable that the private sector is not immune from such experiences, as the now infamous events surrounding the introduction of the London Stock Exchange's TAURUS system testifies (McRae 1993). Examples such as these emphasize the importance of extending the scope of research on the diffusion of innovations beyond the point of adoption. The centrality of implementation to this research makes it worth outlining briefly the concepts that will be developed more fully in the second part of the book.

Implementation

Innovations depend on the process of implementation to become absorbed into the organizational context in which they have been introduced. As a result, if the extent of adoption provides a benchmark as to whether users accept a new product to have potential, implementation determines whether it is utilized. Traditionally, implementation has been envisaged as one phase within a linear progression from initial awareness about an innovation through to routine use. In common with the more general process of diffusion, such an approach increasingly appears to be an oversimplification. For instance, the implementation may take place repeatedly in different sections of an organization, or in some cases even within the same organizational unit. However, the critical feature of this whole process is the difficult task of managing change.

The introduction of an innovation into an organization necessarily implies change. Implementation is a means through which such adaptations are transmitted to the often wary members of the organization. Interpretations as to the significance and complexity of this process vary. Overall, there appear to be three main groups of perspectives. First, *technological determinism* is based on the premise that the benefits of the innovation are so transparent to all concerned that securing utilization is inevitable and largely unproblematic. The second perspective, *managerial rationalism*, is similarly optimistic about the probability of realizing considerable benefits from the introduction of an innovation but instead of relating this solely to the technical characteristics of the new development, the rational application of appropriate management techniques is regarded as making a significant contribution. In contrast, the *social interactionist* perspective suggests that the chances of achieving utilization are far more problematic, with the outcome resulting from the interrelationships between the innovation and the context in which it is located. As a result, objective evaluations of the changes made necessary by the introduction of the new technology or the scope of the benefits are regarded as largely irrelevant to the likelihood of securing satisfactory implementation. Much more important is each individual's assessment of the potential threat or opportunities posed by the new development. In most cases such an appraisal is based on short term personal considerations.

It is increasingly evident that, whether or not the information age has arrived, great significance is being attached to the importance of innovativeness and more particularly the introduction of innovations. Innovations cannot, however, diffuse in isolation and it is therefore vital that consideration is given to improving understanding of the interrelationships between new technological developments and the organizational contexts that they are intended to benefit. The subsequent study will therefore develop and explore the concepts outlined above in relation to the diffusion and most particularly the implementation of geographic information systems in British local government.

Overview

The subsequent discussion is structured around three further parts. Chapters 2, 3 and 4 develop the theoretical framework upon which the study is based. Chapter 2 explores the nature of innovations and their relationship to technological developments such as GIS, and reviews current thinking about the nature of organizations. The impact of the interrelationships between innovations and organizations on implementation provides the focus of Chapter 3. The discussion develops and extends the three perspectives that highlight the varying significance and complexity of this interaction. Having established the theoretical framework for the investigation, Chapter 4 provides a link between the conceptual and empirical parts of the discussion. The underlying research methodology is outlined and key features of British local government described.

Against this background Chapters 5, 6, and 7 present the findings of the investigations examining the diffusion of GIS in local government. Chapter 5 focuses on the extent of GIS adoption. The organizational and technical characteristics of the systems being introduced are also reviewed in Chapter 5, with a threefold typology developed of the styles of implementation that are being applied in the authorities. This typology provides a framework against which the detailed case study findings are presented in Chapters 6 and 7. The central themes of the study are explored in these chapters. Chapter 6 examines the reasons underlying the decisions to invest in the technology and the extent to which the basic form of an innovation is influenced by the organizational context and assesses the levels of GIS utilization taking place. The striking feature of these findings is the limited impact that GIS appear to be having. It is the search for an explanation of the problematic nature of the implementation process that provides the focus of Chapter 7. The implications of these findings have both theoretical and practical dimensions. Chapter 8 draws these issues together, highlighting that the effective implementation of an innovation is highly dependent on the organizational context in which it is located.

CHAPTER 2

Technology and organizations

Introduction

Technologies such as GIS are distinguished by the propensity for organizations rather than individuals to provide the focus for decisions concerning diffusion. As a result the relationship between technological innovations and organizations is likely to have a significant, and even determining, influence on the chances of initially securing adoption as well as achieving long-term utilization. However, in order to understand the interactions between technologies and the organizations in which they are to be embedded it is important to consider the nature of these concepts. Much of the existing literature on GIS treats the technology and the environments in which these systems are expected to make a contribution as largely unproblematic. This discussion will draw on work that has examined both the historical developments of innovations as well as the diffusion and implementation of computer-based systems in general, to assess whether such narrow conceptualizations are appropriate.

The subsequent discussion is divided into three parts. The first examines the nature of technology and innovations while the second explores the characteristics of organizations. In the light of these considerations the final section introduces the notion of reinvention, which questions whether it is appropriate or indeed helpful to conceptualize GIS as one technology or several.

Technology

The term 'technology', like many others, is both highly evocative as well as notoriously difficult to define. Common usage tends to associate technologies with progress and in some ways a near-mystical capacity to improve the future well-being of the whole of society. It is vital, therefore, that the nature of the concept of technology is addressed directly so as to expose the underlying assumptions. Moreover, it is also important to examine the ambiguities that exist between terms such as 'technology', 'innovation' and 'computer-based systems'.

There is a tendency to feel that we can all identify a technology when we see one, although there is often little overlap between these perceptions. This lack of consensus is exemplified by the variety of definitions adopted by the contributors to Goodman et al.'s (1990) investigation examining the interaction between a variety of technologies and the general environments in which they are located. Despite the variability in emphasis it is possible to identify three common elements in any technology: namely machines, methods and knowledge. As a result technology is defined as 'knowledge of cause-and-effect relationships embedded in machines and methods' (Sproull and Goodman 1990, p. 255). An important feature of this definition is that it goes beyond the notion of technology as simply items of equipment. It emphasizes that fundamental to any technology are the underlying techniques and assumptions that are inherent within the methods that a particular set of equipment makes available. Furthermore, this definition suggests that an appreciation of the role and value of these machines and methods does not exist in a vacuum but rather is shaped by existing knowledge. This understanding may be the result of direct experience or, perhaps more importantly, accepted professional or folk wisdom. As a result, while computer-based systems that can store and manipulate map-based data and their attributes may be termed 'GIS', individual perceptions of the nature and utility of that particular technology are likely to vary considerably according to personal, organizational and cultural circumstances.

Danziger, Dutton, Kling and Kraemer (1982) came to a similar conclusion to Sproull and Goodman in a series of studies investigating the implementation of computer-based systems in local government in the United States. They conceptualized the subset of technologies based on computers as a 'package' including hardware and software, people and techniques. In some senses the term 'package' is unfortunate as it has a tendency to become confused with the much more specific notion of a software package. However, the aim again is to place the stress on technology as embracing more than just items of machinery, with the last element referring to aspects such as accepted practices, existing procedures and corporate expectations. King (1985), in a discussion that develops the concept of the information technology package, terms this element 'policy' rather than techniques. He also includes within the package what is termed the broader societal 'infrastructure', which refers to the organizations and networks that influence the development of computing capabilities within a particular context. Like Goodman and Sproull, he also recognizes that any technology is not an independent entity but rather embedded within and influenced by the human, organizational and cultural context in which it is located.

Implicit within much of the discussion about technology is the idea that it is new or innovative. In practice the terms 'technology' and 'innovation' have become virtually synonymous. It is, however, important to distinguish between innovations as 'things' and innovation or innovativeness as a 'process'. One does not necessarily imply the other. For instance, word processors may be regarded as technological innovations although the underlying function that they are fulfilling is much the same as the typewriter or the printing press. Moreover, the extent to which any new capabilities are exploited will depend on the perceptions of prospective users. It is their actions that turn an innovation into something that contributes to an innovative working environment. Innovations are not themselves a source of creativity or change. It is therefore important not to confuse new items of equipment with the capacity to innovate. The most profound developments are usually dependent upon innovations in ways of thinking rather than the introduction of new equipment, although it is quite possible that the latter may contribute to the realization of the former. This in turn challenges notions of innovations or technologies as neutral tools, as a capacity to achieve good or ill is not simply a reflection of the potential offered by these facilities. Issues concerning who will control the technology and what interests will be served therefore become fundamental to the eventual consequences of adoption and implementation. Some would even suggest that ethics and values are not external to the technology but inherent within it (Ladd 1991; Mouritsen and Bjørn-Andersen 1991).

If we return to the definition of technology as consisting of three elements, that is machines, methods and knowledge, it is generally the machines or methods that are new, while the way in which the technology is conceptualized, understood and utilized is dependent upon existing knowledge and practices. Very often the machines may be different but the underlying methods remain the same. Weick (1990) makes a useful distinction between technology and, more particularly, machines which may be fundamentally new, and the technical systems into which this technology is to be embedded which tend not to be. This in turn suggests that for a technology to become widely diffused it must perpetuate the existing trends and even fashions within a particular organization or society. As a result the term 'innovation' is often applied to technologies that may only offer one new dimension. For instance, the machine may be new but the methods and knowledge remain little changed, as is the case with word processing. The degree to which a particular technology may truly be said to be innovative therefore varies greatly.

It is, however, important, as Postman (1992) points out, that the expectations induced by the potential of new equipment do not deflect attention away from asking harsh questions of the merits and value of the facilities that are offered. Very often in practice there is an implicit association between technological innovations and success. Pinch and Bijker (1987) note the findings of Staudenmaier's survey, which analysed the content of 25 volumes of *Technology and Culture* but found only nine articles dedicated to the study of failed innovations. Consequently, much of what may be termed the folklore surrounding technology evokes images of progressive successful environments while those that cast doubt on the appropriateness of such views are regarded as backward and even killjoys.

An important subset of technologies are computer-based systems. Despite the long availability of computing facilities, it is very often the application of these capabilities to a new field that renders the technology innovative. Computer-based technologies are distinguished from mechanical technologies by their programmable nature. This quality enables this specialist set of technologies to be continuously redesigned (Sproull and Goodman 1990). This implies that computer-based technologies may be modified and developed according to changing circumstances. However, as the earlier discussion has demonstrated, the computer is only one element of the technology; equally important are the underlying methods inherent within the system and the existing knowledge which enables sense to be made of the potential offered.

Technology, like innovation, is a highly complex concept. Frequently the terms are used simply to refer to items of equipment. This undoubtedly provides simplicity but this partial view diminishes the chances of understanding the interrelations between a particular technology and the range of environments in which it is embedded and hence the processes influencing diffusion. Given this context, it is important to consider the implications that this view has for the group of technologies referred to as geographic information systems.

Implications for GIS

Implicit within the discussion has been the assumption that GIS should be regarded as a form of technology and perhaps more particularly an innovative technology. There would seem to be considerable support for this notion as GIS combine the key elements of machines, methods and knowledge. In terms of a more detailed categorization, GIS form part of the specialist group of programmable technologies. In fact, like many other technologies at present, the innovative character of GIS stems largely from the most recent developments in computing capabilities. The ability to handle spatial data using computers is by no means new as technologies of this type have been available since the 1960s, while manual forms of such systems have been in use for many decades and arguably centuries before this. However, it is the increased processing capability and resulting speed with which computers can manipulate the vast data sets associated with geographical information that has made GIS a commercially attractive product.

It is not uncommon for developments in a related field to prompt innovation in an area that has remained dormant for some time. However, in considering the innovative character of GIS technology, it is the machine component on which emphasis should be placed. The application of computer technology has not altered the basic concept of a map which remains a symbolic and perhaps even sanitized version of reality. Moreover, the underlying analytical methods such as the ability to overlay various data sets or to construct buffers around key features are by no means new. It is the ease and speed with which these operations may be undertaken that is the facility gained by harnessing the processing capabilities of computers. At the same time the black-box quality of computers tends to obscure the critical assumptions and flaws of the methods which were perhaps all too obvious when utilizing manual techniques. Finally, the existing knowledge through which individuals make sense of the technology with which they are presented remains little altered by the development of GIS.

GIS technologies are therefore innovative to the extent that they make use of the power of computers to process geographical information. However, while GIS have been designed to handle a particular subgroup of data, they are essentially just another computer-based system. Rule and Attewell (1991) suggest that computer-based applications take one of three forms: first, systems that store information and reproduce it in the same form; second, those involving some basic arithmetic operations such as inventory control systems; and finally, the higher-level strategic systems which have decision rules built into their operations. This classification highlights the potential for confusion in dealing with GIS, as inherent within the software sold under this label are all three types of application. The presence of multiple functions within the same product is not, however, unique to GIS. Consequently, it has been suggested that in terms of diffusion GIS should not be regarded as different from other computer-based technologies, most particularly facilities such as management information systems (Eason 1993; Goodman 1993; Moore 1993).

Understanding technology and therefore GIS as not simply items of equipment extends the scope of the issues that will need to be considered in relation to the process of diffusion. Moreover it has been stressed throughout that technologies are not discrete entities remote from the environments in which they are located. The next section will therefore explore the nature of organizations as it is within this context that the critical decisions concerning diffusion take place.

Organizations

The contexts into which computer-based information systems such as GIS are to operate seldom form a significant area of investigation for those involved in the design of these technologies. Even more surprisingly, such an approach often typifies attitudes towards implementation. Technical know-how is frequently valued more highly than organizational sensitivity. This reflects an underlying assumption that the environments into which systems such as GIS are to be introduced all share common characteristics, in particular that rational procedures dominate their daily work practices. Consequently, the setting in which the implementation of a new technology will take place is perceived to be largely unproblematic, while the challenging part of the process is to make the system technically operational. This sentiment has tended to dominate thinking within the GIS field as much as information technology in general (Eason 1988;

Hirschheim 1985). It is, therefore, vital to explore the appropriateness of such assumptions by examining the nature of organizations. The following comments by no means represent an exhaustive review of the literature on organizational theory; however, they are an attempt to identify the key issues and thereby illuminate the subsequent investigation into the implementation and diffusion of GIS in practice. Before examining the nature and characteristics of organizations it is first necessary to consider why it is important to focus on organizations rather than, for instance, the broader context as the key unit of study.

The term 'context' is highly ambiguous and is often used interchangeably with words such as 'environment', 'setting' and 'culture'. Encapsulated within these terms are a vast array of more or less tangible considerations as well as a wide variety of perspectives as to the level of significance with which such attributes should be held responsible for the outcome of the introduction of a new technology. These issues will be explored in the next chapter, but it is important at this stage to examine a further dimension of context, namely the scale at which these attributes are considered. The term 'context' can, for instance, refer to any size of groupings, from a small work unit within a large company to an organization such as a local authority or private corporation, through to a nation-state or even an agglomeration of countries such as the Western world or the Pacific Rim. Each of these have clear institutional boundaries but in addition it is important to distinguish a further element of the wider context, that being professional associations. Such associations tend to exert a significant influence over their members resulting in, for example, individual surgeons or surveyors feeling that they have more in common with fellow professionals than those within the same organization or perhaps country. Similar pressures could also be exerted by ties to family or friends.

In terms of GIS, despite the influence that the socialization within a particular professional culture or personal set of circumstances may have, it is within organizations that these cumulative values take their effect. Organizations come at the intersection between the individuals and the wider context outlined above. They act as the arena in which the culturally influenced or, some might say, determined values and norms of individuals and professional groupings are played out. Sproull and Goodman (1990) endorse this sentiment, arguing that it is at the intersection between levels that the greatest insights can be gained. Crucially, not only are organizations the environment into which individuals bring their own personal set of values, but there is considerable evidence that through the intersection of elements within this context organizations themselves develop their own cultures (Argyris and Schon 1978; Deal and Kennedy 1982; Handy 1991 and 1993; Kanter 1983; Morgan 1989; Schein 1985; Wright 1994). Organizations are therefore rather more than the sum of their parts. As a result, Mouritsen and Bjørn-Andersen have concluded:

> understanding systems development and use involves understanding the relationships between organizational participants and groups. Very importantly, this involves understanding the micro-politics of any social system that is reproduced

by differentially motivated agents and the broader social contradictions that will contextualize the forms of agreement that can be achieved.
(Mouritsen and Bjørn-Andersen 1991, p. 312)

The organizational context therefore appears to be the most appropriate unit of analysis. In practical terms the decision processes that determine whether to acquire a GIS are embedded within organizations. Moreover there is also considerable theoretical support as it is within this arena that the full range of values from personal to global come together, while at the same time each organization tends to develop a personality of its own. Given the significance of organizations, it is now important to explore the nature of these entities in more detail.

Organizations are, as Handy puts it, 'first and foremost, fascinating collections of people' (Handy 1993, p. 23). The existence of organizations reflects the impossibility of a single individual being able to undertake the vast majority of tasks and activities demanded by societies. Moreover, in an effort to achieve their underlying *raison d'être*, organizations develop formal sets of rules as well as symbolic rituals to guide the actions of their constituent members. As a result they are structured groupings rather than *ad hoc* collections of individuals. This is not to suggest that these structures or sets of rules are static: they evolve and develop over time. Organizations have come to pervade all aspects of life. With very few exceptions the production and distribution of goods and services within the private, public and voluntary sectors are conducted through organizations. This is by no means a new state of affairs as organizations have been in existence for thousands of years. Such groupings were quite familiar to the ancient Greeks, for example, and there seems no reason to assume that they will not continue to have a significant role.

There is little dispute over the most basic features of organizations described above. However, closer examination of the nature of organizations has resulted in profound variations in emphasis and widely different interpretations of the underlying processes influencing the chances of securing effective daily operation. The diversity of approaches is well illustrated by the eight metaphors developed by Morgan (1986). He suggests that organizations can be understood as machines, organisms, brains, cultures, political systems, psychic prisons, flux and transformation and instruments of domination. Inherent within each are very particular sets of assumptions about the nature and purposes of organizations and consequently widely different prescriptions as to the most appropriate actions to take in order to ensure organizational survival. There are no right answers and in many cases individuals within the same context may have very different perceptions of the nature of the environment in which they are located. Moreover, in theory such states are by no means constant, although as we shall see later most organizations tend to be resilient to change. Handy (1993) emphasizes in his seminal work on understanding organizations that the overriding characteristic of organizations is the absence of predictive certainties. This is attributed to the vast array of variables that impinge on any one organization, while at the same time the individuals found within these environments show an inherent ability to defy expectations.

Such complex and yet at the same time important contexts for the economic, social and perhaps even psychological well-being of societies and individuals, have prompted a plethora of studies and theories. In terms of the diffusion of information technologies there appear to be two key elements that need to be explored. These are styles of bureaucracy and approaches to decision-making. These two aspects are closely related, perhaps even interrelated, but it is useful to distinguish one from the other for the sake of clarity. The first refers to the norms and values of the organization which are reflected in, for instance, routine practices, general expectations, styles of leadership and the staffing structure, while the second focuses on the formal and informal procedures for decision-making and in particular the role of information in this process.

The image of organizations as machines has tended to typify the underlying assumptions of both those involved with the development of computer-based systems in general as well as GIS. This approach is dominated by sets of rules which are designed to produce highly ordered and efficient work environments and is most closely linked to Taylor's work on scientific management (Taylor 1947). In many ways this perspective represents an ideal of how organizations should operate in the absence of people and an ever-changing context. As a result there is a tendency to suggest a uniformity of approach, that is to say that for all activities there is an optimal style of organizational working. The task for the manager is to identify the appropriate strategies with the inherent logic of the approach, resulting in its inevitable implementation. It is implicit within such a conceptualization that there is a shared understanding among all staff as to the goals of the organization and widespread consensus as to their efficacy. The importance of formal and rationally based rules as the foundation for legitimate authority in organizations is also stressed in the work of Weber (1947) on bureaucracy. These rules and routine procedures were regarded as providing consistency and predictability and, by de-personalizing administration, were seen as superior to reliance on the charismatic qualities of a single individual or the blind adherence to historical precedent. The result is highly ordered hierarchical organizations with clearly specified goals. As Haynes puts it, 'For Weber the ideal bureaucratic organisation was essentially the well designed machine' (Haynes 1980, p. 8). Given their perspectives it is perhaps not surprising that the process of securing the effective utilization of information technology is regarded by those who share these assumptions as largely unproblematic.

These normative explanations of how organizations should operate provide the basis through which subsequent perspectives depart. The concept of an organization as an organism introduces an emphasis on the determining influence of the external environment. The scope of action of individuals is therefore strongly affected by events outside the organization. There is also an underlying sense that Darwin's ideas of the survival of the fittest applies just as much to organizations as to the wide range of biological communities. A link from this can be seen in later work which applies open-systems theory to organizations. An organization under this set of theories is viewed as an open system which takes in resources from the external environment and processes and transforms them into a new product which is then returned to the wider context. The key feature is the mutual dependence between the organization and the environment in which it is located so that everything within this system affects everything else. As a result the models that describe organizations in this way tend to be highly complex.

Implicit within conceptualizations of organizations as machines, organisms and systems is an underlying assumption that individuals within such entities share a common purpose. This view has, however, been challenged. Much evidence now suggests organizations to consist of coalitions which spend much of their time in competition, if not in outright conflict. As Hirschheim states, 'Offices are not rational and manifestly rule following, they are social arenas where power, ritual and myth predominate' (Hirschheim 1985, p. 279). Each organization is therefore viewed as a unique social system in which individual members are socialized in a particular set of norms, beliefs and values (Schein 1980). As occurs within societies, there will be disputes over goals and priorities but the attitudes towards such non-conformist behaviour is a reflection of much more deep-rooted characteristics, often referred to as the organizational culture. Consequently, an organization is viewed as much more than an *ad hoc* collection of individuals. The constituent members are at once unpredictable in their behaviour and yet strongly influenced by what Deal and Kennedy (1982) in their work on organizational culture described as 'the way we do things around here'. The implication is that there is an unconscious set of shared beliefs which guide the conduct of individuals within a particular organizational context (Schein 1985). In most cases these values and norms are not specified but are just part of the procedures and rituals of the organization. Argyris and Schon (1978) actually distinguish between the public image that organizations seek to promote, as expressed in, for instance, formal documentation and the reality of daily practice. The former is referred to as 'espoused theory' and the latter 'theory-in-use' and they note that there are frequently inconsistencies between the two.

The term 'culture' with reference to organizations encapsulates a vast array of considerations both in and around organizations (Frissen 1989). These vary from the tangible, such as the level of financial resources available, skills of the workforce and institutional arrangements, to the more abstract, for instance the values, beliefs and motivations of individuals, the scope of ethical responsibility and attitudes towards morality including such issues as privacy and confidentiality. There is, however, a lack of consensus as to the degree of emphasis that should be given to the cultural aspects of organizations. Frissen distinguishes four approaches as follows:

1. *Culture as a contingency factor.* Under this conceptualization a systematic relationship is envisaged between different variables whereby variation in

any one element has implications for the rest (see, for instance, Child 1985; Hofstede 1980).

2. *Culture as subsystem.* This view suggests that culture exists as a separate subsystem which can be distinguished from others such as work process or management structure. Issues within the cultural subsystem include risks, rituals, values, norms, symbols, relationships between individuals, acceptable patterns of behaviour and leadership, most particularly how control should be exerted (see, for instance, Deal and Kennedy 1982; Peters and Waterman 1982).

3. *Culture as aspect system.* All subsystems are viewed under this approach as having a cultural dimension. For instance, managerial structure is not just a formal assignment of roles but embodies the values and norms associated with these tasks. (See, for instance, March and Olsen 1983; Meyer and Rowan 1977.)

4. *Organization as a cultural phenomenon.* The central proposition of this view is that an organization, 'does not have a culture, but it is a culture' (Frissen 1989, p. 572). Culture is seen as the very essence of an organization rather than one, although perhaps highly significant, aspect. Weber (1947) regarded it as impossible to separate an organization from its cultural context as they are one and the same.

Current discussions based on the concept of organizations as cultures most closely resemble Frissen's categories of culture as either a subsystem or an aspect system. The study of organizations, therefore, becomes in many ways an historical analysis: a process of uncovering the traditions and rituals that have resulted in a particular style of working. Work by Harrison (1972), which has subsequently been developed by Handy (1993), emphasizes that while each organization should be regarded as unique, it is possible to distinguish four dominant cultural types. These are as follows: *power cultures*, which tend to be dominated by an all powerful head and are characterized by an absence of formal structures and a willingness to follow the lead of the charismatic central figure; *role cultures*, which in contrast are based on logic and rationality and are dominated by rules and procedures and as such are typified by conformity and predictability; *task cultures*, which tend to be team-based environments and place an emphasis on getting things done without strong control from the top; and finally *person cultures*, which are regarded as existing to serve the individuals and, as such, organizational goals are secondary to the desires of the individual.

The value of this classification lies not so much in the extent to which a particular organization can be said to conform to one of these conceptualizations but in the massive variations in styles of working that they suggest. It is implicit within this that individual organizational cultures will have highly distinctive attitudes and approaches to innovation and more particularly the role and value of computer-based systems such as GIS. Some, such as the archetypal task culture, may be happy to embrace change and demonstrate an ability to cope with and perhaps even thrive on disruption while others, for instance role cultures, exhibit a reticence and scepticism about the value to be gained from diverting resources to activities which may in the end prove fruitless. This is not to suggest that there is one superior culture which all organizations or even companies within a single industrial sector should seek to achieve. The 'organizations as cultures' perspective suggests that every organization has its own way of doing things based on its own history and traditions as well as the manner in which external pressures and circumstances have been internalized. However, at the same time it is acknowledged that this approach is not necessarily the best or the only way to undertake their particular tasks (Handy 1993). Consequently, the nature of the issues that need to be considered in relation to the implementation of GIS and the propensity for effective utilization to be achieved differ markedly from the more mechanistic conceptualizations.

A development of the notion of organizations as cultures is to suggest that, rather than being simply social arenas, organizations are an institutional device for political as well social domination of the majority by the most powerful within society. Pfeffer (1981 and 1992) suggests that organizations are characterized by the interplay of power and politics in which the objectives of individuals or coalitions are frequently in conflict. However, these groupings are not equally powerful and as a result it is argued that those in the most favourable positions use this advantage to strengthen their situation. In a sense, therefore, organizations are simply regarded as a microcosm of the more general processes acting on the whole of capitalist societies. Who governs the organization and who benefits from the decisions that are being made become the crucial issues for analysis. The introduction of new computer-based technology is considered to be part of the same process, whereby the most powerful grouping within the organization gains the greatest benefit. The outcome is therefore predetermined by powerful structures within society. Giddens (1979) has developed this further, viewing structure and actions as a duality. As a result it is suggested that organizational practices and structures are shaped by the actions of individuals within these environments which in turn reaffirm or in some cases modify these practices and structures. This perspective to some extent links notions of organizations as arenas for domination with that of cultures by suggesting that individuals within such contexts have the capacity to decide whether or not to perpetuate the existing order. It therefore follows that organizations may well take a wide variety of forms.

A closely related issue to the style of working within organizations is the nature of the decision-making process. Given that the justification for technologies such as GIS is the contribution that they will make to operational, managerial and strategic decision-making, it is important to explore the underlying assumptions about this activity. The range of perspectives reflects a similar set of theoretical positions to that concerning the styles of working in organizations. As a result Table 2.1 takes four of the key organizational metaphors discussed above and outlines the main characteristics of decision-

GIS and organizations

Table 2.1 Key characteristics of a selection of theoretical approaches to decision-making in organizations

Characteristic	Organizational metaphors			
	Machine	Bureaucracy	Culture	Instruments of domination
Decision-making style	Rational Orderly Comprehensive Optimal	Procedural Rational Orderly	Disorderly Compromise Negotiation and bargaining	Confrontational Predetermined
Decision-making process	Problem → information → decision	Problem → information → decision, guided by rules and procedures	Ill-defined problem → search for information based on agreed values → compromise	Decision → propaganda → conformity
Role of information	Substantive	Substantive	Symbolic Ritualistic	Symbolic Political
Role of computer-based data	Increases rationality	Increases rationality	No different to other sorts of data and less important than shared understandings	No different to other sorts of data
Information and computational requirements	Extensive	Constrained by rules	Dependent upon organizational norms and values	Information used and withheld in order to maintain control

making that are generally associated with these perspectives. This is not designed to be a comprehensive review of the literature on decision-making but rather to highlight a number of key issues. As a result particular focus is placed on the role and significance of information in general and especially computer-based information in this process.

The style of decision-making associated with the metaphor of organizations as machines is assumed to be highly rational and orderly in nature. The process is regarded as linear, with the realization of a problem resulting in the search for information which in turn determines the most appropriate course of action. Information therefore performs the pivotal role in the process, providing the necessary evidence to indicate the optimal solution to whatever dilemma is facing the organization. While not challenging the appropriateness of aiming for such an approach, studies by Simon (1952), for instance, suggested that the actual ability to find an optimal solution is constrained by the skills of the individuals involved and the availability of appropriate information. Given such an analysis as to the main weaknesses within the decision-making process, it has been assumed that by harnessing the power of computers it will be possible to compensate for individual failings and at the same time increase access to information and reduce the opportunity for details to be filtered or distorted by the staff responsible (Simon 1973; Whisler 1970). The justifications for the adoption of GIS technologies tend to reflect this type of thinking.

A similar conception of decision-making as a formal–rational process dominates the assumptions underlying approaches that suggest the most appropriate organizational form to be a bureaucracy. Weber (1947) stresses that rationally based decision-making is the only basis for legitimate authority. However, in order to achieve what he regards as the highest form of decision-making, comprehensive sets of rules and procedures are necessary to guide the process. It is within this context that computerization is regarded as having a significant contribution to make as such systems reinforce existing procedures and thereby lead to better quality decisions.

The underlying assumptions of these essentially normative views of decision-making are challenged by the perspectives that envisage organizations as either cultures or instruments of domination. In the case of the former, selecting the most appropriate course of action is conceived as a complex interactive and even disorderly process. Decision-making in such circumstances is regarded as a process of negotiation and compromise with the initial problem likely to be very poorly defined and the eventual decision dependent upon achieving consensus based on a set of agreed values. Information is therefore deployed as ammunition within the bargaining process and is used to justify the final compromise. Consequently, information is regarded as just one source of evidence. Equally important are other forms of knowledge such as experience, precedent, beliefs, values, gossip and social status. Support for this perspective has been provided by research undertaken in organizations responsible for such diverse activities as mental health and accountancy (Argyris 1971; Feldman and March 1981; Larsen 1985; March 1987; Weiss 1977). Feldman and March's work, for instance, suggests that information has a symbolic rather than substantive role. The process of data collection and analysis is regarded as signalling to other organizations that decision-making has been undertaken in a rigorous and systematic manner. Moreover arguments that are supported by information symbolize rationality and individual competence, while the initial decision is often a result of intuition and emotion rather than objectivity and rationality (see also Argyris 1971). Computer-based data are regarded as little different from information emanating from other sources. They perform the same symbolic function with the precise nature of the role depending on the values, norms and rituals of the organization. It follows, therefore, that the sort of evidence that tips the balance of the argument in one direction rather than another is perceived to vary between organizations according to their traditions and values.

The view of organizations as instruments of domination suggests that while conflict and bargaining typify decision-making in organizations, the outcome of this process will inevitably favour the most powerful coalition or individual. As a result it is envisaged that the decision is formulated prior to the search for supporting evidence, with the information collected used to convince dissenting interests. Information derived from computers is regarded as equally subject to selectivity, distortion and manipulation by the dominant grouping. However, the symbolic value of the association between computers and rationality is assumed to favour those in the most powerful position as they control resources. Computerization, it is argued, may even strengthen their position, assisting them in the process of generating information in order to support their interests (Kling and Iacono 1984; Markus 1983; Pfeffer 1978 and 1981). A variation on this line of thinking suggests that the introduction of computers will lead to a shift in power towards the technical specialists as they will act as the gate-keepers of a vast store of information (Downs 1967; Markus and Bjørn-Andersen 1987).

Summary

The wide variation in underlying assumptions with respect to the nature of organizations and the key features of the decision-making process within these environments has significant implications for the diffusion of GIS. Much of the existing work on the diffusion and most particularly the implementation of this technology is based on normative explanations of how organizations *ought* to operate. However, analyses conducted in a range of environments challenge the appropriateness of such approaches and therefore the resulting prescriptions as to the method and ease of implementation as well as the role performed by such technologies and value of the information generated. As Handy states:

> Organisations are not machines, even though some of those running them would dearly like them to be so. They are communities of people, and therefore behave just like other communities. They compete amongst themselves for power and resources, there are differences of opinion and of values, conflicts of priorities and of goals. There are those who want to change things and those who would willingly settle for a quiet life. There are pressure groups and lobbies, cliques and cabals, rivalries and contests, clashes of personality and bonds of alliance. It would be odd if it were not so, and foolish of anyone to pretend that in some ideal world those differences would not exist.
>
> (Handy 1993, p. 291)

It is in the light of such evaluations that the subsequent investigation is undertaken. Organizations are not simply machines and to assume such is to ignore a set of social and political processes which at the very least make a significant contribution to technological diffusion and perhaps even determine the outcome. There seems little reason to suggest that the set of organizations that provide the focus for this study, namely local authorities, behave any differently than organizations in general. Chapter 4 will explore the nature of British local government in detail but at their most basic they consist, like organizations in the private and voluntary sectors, of people with sets of tasks to perform. These tasks are many and varied and in turn are dominated as in other environments by a variety of professional interests. Despite externally enforced changes the stress in local authorities is still on effective service delivery rather than profit generation. However, this does not appear to affect significantly the nature of the processes at work within organizations.

The significance of social and political processes in understanding the nature of technology and organizations has led to suggestions that innovations are not just invented but constantly reinvented. The final part of this chapter will therefore explore the nature of reinvention.

Reinvention

The review of the nature of technology emphasized that just as important as the precise technical configuration of the innovation are the perceptions that individuals develop as to its role and value. In turn the analysis of organizations suggests that these perceptions are likely to vary according to the culture of the particular contextual setting. As a result two organizations may purchase exactly the same items of equipment but their understanding as to the nature of what they have acquired is likely to be entirely different. To complicate matters further, similar discrepancies may be found between the views and expectations of individuals within the same organization. Consequently, Rogers (1993) has pointed to the propensity for the same technology to be repeatedly invented in a variety of forms in different organizational settings. Reinvention is therefore viewed as an increasingly significant part of the diffusion process. It would be inappropriate to castigate such activities as simply ill-informed individuals or poorly managed organizations wasting resources by 'reinventing the wheel'. It should more properly be seen as a process of reshaping the wheel so that it actually fits the dimensions and perhaps even aesthetics of the particular organizational cart to which it is to be fixed.

The capacity for reinvention to take place is likely to have significant implications for the study of the diffusion of GIS technologies. It suggests that while the label 'GIS' may be linked to the product purchased by an organization, perceptions as to its nature and value may in practice vary significantly between different settings. It is evident that even within the academic community there are considerable differences in the accepted technical definitions of what constitutes a GIS. Such ambiguity is further intensified when the views of potential or actual users are included. These individuals tend not to regard GIS as a specific collection of computer functionality; rather they see it in terms of applications, output and, importantly, caricatures of other people's experiences. In a sense, therefore, seeing is believing and each individual's or, at a collective level, each organization's interaction with an innovation such as GIS is a process of discovery and learning: a process, however, that does not take place within a vacuum and may well be shaped by existing traditions, beliefs and prejudices.

Conclusion

The overview of the literature on technology and organizations casts considerable doubt on the appropriateness of simplistic notions that either of these concepts is analogous to a machine. The complexity of the issues involved and, more importantly, the role of social and political processes has significant implications for the nature of the interrelationships between a newly acquired technology and the organizational context in which it is to be embedded. At the most basic level these considerations point to the importance of exploring not only the technological aspects of diffusion but also, and perhaps more importantly, the underlying organizational processes. In the light of this evaluation the following chapter will explore a range of perspectives on the implementation of GIS.

CHAPTER 3

Perspectives on implementation

Introduction

The process of implementation is critical to the diffusion of any technological innovation as it is this process that is responsible for transforming unproven potential into a taken-for-granted component of the daily activities of an organization. As the discussion in the opening chapter pointed out, implementation is not regarded in this study as one stage within an inevitable linear progression towards utilization. In practice it can be very difficult to determine the start or finish of implementation as important decisions are often negotiated long before the actual purchase of the equipment or facility. Moreover, in the case particularly of programmable technologies it is rare for there to be a single phase of implementation. Much more typical appears to be repeated cycles of design, learning and use with each cycle reflecting changing circumstances, growing understanding and the vagaries of negotiations between interested parties. Implementation is not therefore conceptualized as simply a stage on the path towards routine use but rather a complex and somewhat problematic process.

The implementation of GIS technologies would seem to pose two key questions for organizations: first, how to integrate a new working practice into the existing traditions and norms, and second, in what way will the information that is generated contribute to the decision-making processes within the organization. Fundamental therefore is the capacity to manage change and most particularly to sustain this process over a considerable time period. The preceding chapter identified such facets of organizational life as crucial to the various perceptions of the nature of organizations. This chapter will develop these ideas further, focusing on the impact that the interaction between organizations and a technology such as GIS has on the process of implementation. Similar issues have been explored in relation to the diffusion of information

GIS and organizations

technologies in general (Campbell 1995). However, it must be emphasized that in this discussion the specific concern is the process of implementation and consequently it is not appropriate to explore broader issues questioning the inherent value of computer-based technologies, such as arguments that challenge the ethical and moral worth of machine-based societies (Postman 1992) or express concern about the contribution that information technologies are making to the creation of increasingly fragmented and unequal societies (Braverman 1974; Howard 1985; Noble 1984). The insights provided by studies investigating the development of computer-based systems in practice suggest that there are three groups of perspectives on the nature of the implementation process. These are technological determinism, managerial rationalism and social interactionism. The subsequent discussion will examine each in turn and consider the implications for the development of GIS.

Technological determinism

The key feature of technological determinism is the stress placed on the inherent technical worth of a particular innovation. This facet is perceived to override all others and as such is sufficient to justify acquisition as well as to lead inevitably to utilization. Consequently, the underlying rationale is based on the premise that the technological advantages of a particular innovation will be so transparent to the members of an organization that they will readily embrace the new approach. Put more simply, if someone develops a better washing machine it is bound to sell and to be used because it is demonstrably better than its competitors. It therefore follows that implementation is regarded as little more than fine-tuning the technology, with the human and organizational component of this process largely overlooked. Table 3.1 provides a summary of the main characteristics of

Table 3.1 Main characteristics of technological determinism

Characteristic	Technological determinism
Propensity for adoption	Inevitable if a good technology
Reason for adoption	To solve an operational problem that has been identified
Style of implementation	Technical process
Constraints on implementation	Technical worth of the innovation and stupidity of users
Likely outcome of implementation	Positive, greater efficiency and more rational decision-making
Perception of technology	Machine and methods
Perception of organizations	Machine, organism

technological determinism. This framework will be used to examine the perspective in more detail.

Technological determinism is essentially based on a utopian view of technical developments (see, for example, Feigenbaum and McCorduck 1991; Naisbitt 1984). As a result, given the inherent technological quality of the innovation it is regarded as virtually inevitable that it will be extensively adopted. Furthermore the acquisition of a technology is generally assumed to imply utilization. It therefore follows that if simply purchasing an innovation is sufficient to ensure use, then implementation must be a relatively trivial and unproblematic process. It is implicit within this understanding that the organizational context into which this technology is to be located is regarded as an irrelevance. In some cases account is taken of the external environment in the sense that economic survival is said to dictate that new working practices are adopted. Overall therefore the process of diffusion is based on a logical association between the inherent value of, for instance, computer-based technologies and optimizing the goals of the organization. Moreover such an objective is regarded as providing the underlying motivation for all work within a particular environment.

These assumptions suggest that the process of adoption and implementation is guided by the inherent value of the technological innovation and is a rational response on the part of the participants within an organization. As a result, while widespread acquisition is assumed to result from the creation of a technically superior product, the motivation for the introduction of the technology by an organization stems from the desire to correct an operational weakness. This suggests that the identification of a difficulty prompts the search for a solution which in turn leads to the purchase of an innovation such as a GIS. In these circumstances the technology would not be introduced into the organization without a particular task having already been identified. Therefore the aim of implementation is to realize this objective and consequently improve organizational effectiveness and efficiency. It is rare for the outcome of this process to be conceptualized as other than positive for all concerned. The achievement of organizational goals or improvement in overall performance are assumed to equate with personal targets. As innovations enable old tasks to be undertaken more effectively as well as open up new areas of activity, it is often simultaneously perceived that working conditions will also be enhanced. Such thinking is exemplified by Giuliano's (1991) evaluation of office work in the information age. He envisages office activities as being transformed so as to become both more flexible and efficient in nature as well as creating more cooperative and interesting working environments; a situation which must be to the mutual benefit of all.

Such analyses suggest that the advantages of computer-based innovations are so transparent that implementation is no more than a technical process. Given that the organizational context is regarded as largely unproblematic, the most important quality is technical competence. Under these circumstances decisions about the most appropriate product to purchase and the overall strategy for system design, including such important issues as customization, typically become the exclusive responsibility of technical specialists. This leads in turn to priority being given to the most advanced and powerful system rather than one that equates with the organizational environment into which it is to be located. It is inevitable that, in circumstances where the implementation of technological innovations is viewed as a purely technical activity, those possessing such skills dominate. This issue is not represented in this perspective as part of a power struggle (see, for example, Downs 1967; Markus and Bjørn-Andersen 1987) but rather the logical approach to adopt, given the technological nature of the activity.

The conceptualization of implementation as essentially technical in nature limits the range of issues that need to be considered. As a result the main constraints to achieving the effective utilization of a new technology are technical failures or the incompetence of potential users. The first acknowledges that in practice innovations do not always fulfil expectations. It is quite possible that under daily operational conditions technical inadequacies may lead to the rejection of a technology. However, this is often viewed as only a temporary setback. The principle behind the product may still be regarded as valid and therefore eventual implementation is simply a matter of further technical refinement or more powerful software as in the case of GIS. Technological developments in related fields are also regarded as making a significant contribution. In relation to GIS, technological advances in the methods of the collection and conversion of geographical information are often seen as a precursor to the effective implementation of such systems. The second set of factors accounting for the failure to employ an innovation is the ignorance and lack of skill of potential users. In this case there is an underlying sense that if only individuals could fully comprehend the potential, they would grasp the opportunities offered. As Kling puts it, referring to the broader ramifications of technological development, 'Perverse or undisciplined people are the main barriers to social reform through computing' (Kling 1991, p. 355).

Technological determinism is imbued with a sense of fatalism about the inevitable progress of automation and optimism about the impact that this trend will have on individual organizations as well as society as a whole. Emphasis is placed on a narrow conceptualization of the technology as a set of machines and methods virtually in isolation from the context in which it is located. Such thinking dominates much of the GIS literature (see, for example, Maguire, Goodchild and Rhind 1991). As a result, stress has been placed on technological developments and improvements in know-how to the virtual exclusion of consideration of broader issues and ramifications. Innes and Simpson's survey of articles and abstracts concerning GIS in the Proceedings of the Urban and Regional Information Systems Association Conference in the United States points to the nearly all-pervasive view that the design of more advanced and, it is assumed, necessarily better GIS technologies will inevitably result in their adoption and utilization. They state: 'Often the articles read as if developing more powerful and user-friendly applications will automatically result in the

Perspectives on implementation

blossoming of GIS in practice' (Innes and Simpson 1993, p. 230). The general literature on GIS reflects this line of thinking. GIS is seen as unlocking a vast array of geographically based data sets. However, while this technical capacity is undisputed, little consideration is given to the purpose to which such a facility would be put or the extent to which users perceive it to be necessary or valuable. The underlying assumption of the technological determinist perspective is that utilization and widespread diffusion of an innovation is dependent upon its technical qualities. Consequently, implementation is conceptualized as a largely technical and relatively straightforward process, which is regarded as leading to the final realization of a long-anticipated set of opportunities which have only been frustrated by the limitations of the currently available technology. Technical viability is therefore perceived to be sufficient to ensure widespread use.

Managerial rationalism

Managerial rationalism departs from the preceding perspective in that the process of introducing an innovation into an organization is not regarded as solely technological in nature. Effective implementation is viewed as a combination of 'good', perhaps more precisely, 'rational' management and technical competence. Innovations are not simply expected to be utilized because of their inherent potential but rather as a consequence of the strategies laid down by the managers within that environment. Table 3.2 outlines the main characteristics of managerial rationalism. This framework will provide the basis for the subsequent discussion.

There is an underlying clarity to managerial rationalism based on the assumption that individuals within organizations act rationally and will follow

Table 3.2 Main characteristics of managerial rationalism

Characteristic	Managerial rationalism
Propensity for adoption	Inevitable if a good technology and rational management
Reason for adoption	To solve an operational problem that has been identified
Style of implementation	Guided by a rational management strategy
Constraints on implementation	Poor management and the technical worth of the innovation
Likely outcome of implementation	Positive, greater efficiency and/or more rational decision-making
Perception of technology	Machine and methods
Perception of organizations	System

the lead taken by senior staff. In common with the previous perspective, personal goals are regarded as secondary to achieving the objectives of the organization. The guiding principles of logic and rationality mirror in many ways Taylor's (1947) views on scientific management and later developments which applied open-systems theory to the operation of organizations. As a result, adoption of an innovation is perceived to be inevitable if it has the potential to enhance efficiency or overcome human limitations in decision-making. It is therefore assumed that acquisition of a new technology will naturally follow from the identification of a weakness in the daily operations of the organization or, more profoundly, strategic activities such as decision-making. In the case of the latter the area of concern is likely to centre on a feeling that senior managers lack sufficient information with which to select optimal courses of action.

Implicit within this conceptualization is a linear view of diffusion, with the identification of a problem logically leading to the search for a solution which in this case is represented by a new work practice. The whole essence of scientific management is pre-planning and strategy formulation: a style of working that, in terms of the introduction of technology into an organization, is represented by a sequence of stages leading inescapably towards the utilization of the system. This is exemplified by what Eason (1988) describes as the traditional data-processing approach. This approach consists of eight separate phases, namely:

1. *project selection* which provides the design team with the terms of reference to be followed;
2. *feasibility study* which specifies the costs, benefits and plan to be followed;
3. *systems analysis* which involves the systems designers in measuring volumes and flows of data;
4. *requirements specification* which identifies the size and nature of the system that will be needed;
5. *systems design* which represents the point at which the detailed technical requirements of the system are defined, such as the database model to be used;
6. *construction* which deals with technical development of the system;
7. *trials* which entail testing the technical performance of the system; and
8. *implementation* which involves distributing the system to users and establishing procedures for its operation, maintenance and support.

This approach defines implementation narrowly as simply involving the specification of rules and procedures while the whole process is dominated by identifying the technical elements of the system rather than the organizational aspects. There is an overwhelming feeling that the outcome of all this pre-planning will be the perfect system on paper but one that might not be appropriate to the organizational environment in which it is located.

Later adaptations of the data-processing approach have acknowledged the need to incorporate users into the overall structure. As a result, structured design methods follow a similar linear sequence but at the end of each stage users are invited to review progress. Eason (1988) has criticized these approaches, for while they include users, their role in practice is highly constrained. In particular he points to the limited time that is often available for users to assimilate and contribute to the complex and highly technical documentation with which they are confronted. This form of consultation is therefore regarded as no more than a token gesture, with the focus of the overall process still concentrating on the technical aspects of implementation rather than the organizational implications.

It is implicit within managerial rationalism that it must be possible to identify one optimal strategy for the implementation of a particular technology which will prove successful in any environment. The only factors that are likely to hamper the realization of the merits of the innovation are technical inadequacies or failure to take account of all the issues involved during the pre-planning phases, particularly omitting or not undertaking each stage in the specified order. There is, therefore, an underlying assumption that all aspects of the implementation process can be specified and furthermore that good intentions on paper will automatically be turned into actions based on their inherent rationality. It is also implicit that conditions will remain largely stable. Given such an analysis it is inevitable that the outcome of this process is perceived to produce widespread benefits. It is assumed that as the purpose of the introduction of the technology is to contribute to the effective realization of the goals of the organization, this must in turn benefit all those within that context as their personal aspirations are regarded as synonymous with those of the organization.

Unlike technological determinism the managerial rationalist perspective acknowledges that the introduction of a new technology needs to be managed and coordinated if it is to yield its full potential to the organization. However, in both cases the basic underlying philosophy centres on rationality, whether this is simply conceived as the way the world operates as in the former or as a guiding management style in the latter. This results in both placing considerable stress on the technical aspects of implementation, although in the case of managerial rationalism within a framework which suggests the importance of pre-planning and the need to consider the organizational arrangements even if this is only to the extent of setting procedures. This in turn suggests that having formulated an optimal plan or strategy it is possible to implement these to the full. Consequently, this approach is often referred to as the cookbook method of system implementation. This implies that if you have all the ingredients and follow the recipe to the full, the result will always be a well risen soufflé or, in this context, a well-utilized GIS. The next section will examine how such thinking has been applied to GIS.

Implications for GIS

It has been emphasized throughout this discussion that the vast majority of the work on GIS focuses on the technological dimensions of these systems assuming to a large extent that enhancements in technical performance will lead to widespread utilization. However, those studies that have explored the process of implementation have tended to base their analyses on a managerial rationalist perspective. Perhaps the largest quantitative contribution to this area has come from those drawing on their management consultancy experience. A scan of the proceedings of the main GIS conferences quickly reveals numerous recipes for success, all based on an essentially rational understanding of the workings of organizations (see, for example, Mahoney and McLaren 1993; McAusland and Summerside 1993). A common element of much of this prescription has been the recommendation that the most effective strategy for GIS implementation is one based on a 'corporate' approach. This is by no means a new prescription as the most effective method of introducing computer-based systems into organizations. It has long held sway with system designers and has now been embraced by much of the GIS industry.

The widely held belief in the advantages of corporate implementation is part of the management science tradition which assumes that organizations are arenas of rational activity where logical strategies are immediately turned into action resulting in greater organizational efficiency. The term 'corporate' is used to imply at the very least a coordinated approach to the introduction of GIS which involves the vast majority of the departments or subsections into which an organization is divided. The same concept is often extended beyond the limits of a single organization and may therefore be used to refer to a multi-agency development. The underlying assumptions of the corporate approach are that computer-based systems yield the greatest benefits and contribute to efficiency when the number of organizational units participating in the project is maximized, with each abiding by a common set of standards and procedures for the adoption, implementation and maintenance of the system.

Implicit within this approach is the need for centralized control either as a result of direct imposition through a top-down management strategy or perhaps in a more cooperative manner through a process of negotiation and agreement. The latter is not regarded as particularly problematic as the benefits to be achieved are perceived to outweigh by far any loss of independence. It is usually assumed that the technical specialists are the group best equipped to lay down or draft the appropriate standards and specifications, with senior managers acting as the final arbiters. A variation on this theme has been a suggestion that a federal rather than corporate approach is likely to prove a more effective strategy (Barr 1991). It is envisaged that this approach to implementation provides individual organizational units with greater flexibility but within an overall framework of standards which ensures compatibility. There are obvious parallels between this organizational structure and the lively discussions over federalism that have recently taken place within the European Union. In particular the term 'subsidiarity' has become synonymous with wide variations of opinions and beliefs, in this case as to the types of issue that are most appropriately handled centrally as against individual nation-states. It would seem likely that federalism within a GIS context would result in the same passionate debate. Despite these

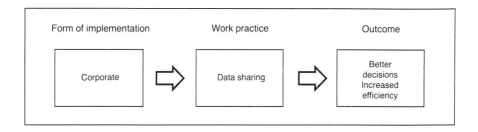

Figure 3.1 Underlying rationale of the corporate approach to GIS implementation

variations in emphasis, the underlying characteristics of a corporate approach are that system implementation should be multi-departmental or even multi-agency in nature, with the overall coordination of the project directed by a common set of standards and procedures which have been established centrally.

The rationale behind corporate implementation is that such a strategy will increase levels of data sharing within an organization and thereby reduce duplication as well as leading to more informed decision-making. Similarly the main advantages of GIS technologies are regarded as the ability to integrate data sets from a wide variety of sources and, through the medium of computers, to make this information more widely accessible as well as available in a variety of new forms. As a result it has been assumed that the most appropriate means of securing these benefits is for GIS to be implemented corporately. The logic of this argument is illustrated in Figure 3.1. This suggests that it is only through a corporate strategy that widespread information sharing will take place and that without data sharing, duplication and poorly informed decisions will result. Such circumstances are in turn likely to threaten the survival of the organization. The corollary to the same line of argument is that the development of separate systems will prevent information sharing as it is probable that both the technology and the data will be incompatible. As a result there will be little possibility of securing the enhancements in organizational working that are assumed to follow from a corporate approach. Moreover, the organizational unit is liable to face considerable isolation and as a consequence will miss out on the resources that are available centrally such as technical specialists, training and even additional funding. Table 3.3 provides a summary of these arguments.

Corporate implementation is therefore often regarded as the only strategy that will yield the strategic benefits and efficiency gains that are embedded in GIS technology (see, for example, Bromley and Selman 1992; Coulson and Bromley 1990; Gault and Peutherer 1989; Mahoney 1989). Furthermore, such an approach is not regarded as a short-term expedient through which resources can be pooled so as to cover the initial costs, but rather as an ongoing commitment. For instance, Gault and Peutherer (1989) have argued in relation to local government in Great Britain that, given the general restraint on spending, the

GIS and organizations

Table 3.3 Arguments in favour of a corporate strategy towards GIS implementation

Advantages of a corporate strategy	Disadvantages of separate system development (a departmental approach)
Integration of formerly separate data sets	Isolation of each organizational unit
Increased capacity for data sharing	Limited support in terms of finance, technical specialists and training
Improved access to information	Incompatible systems and data
More informed decision-making	Continued duplication of effort
Increased efficiency due to reduced duplication	Ill-informed decision-making

introduction of compulsory competitive tendering for services and calls for greater accountability, it is essential that local authorities adopt 'more rational decision-making procedures' (Gault and Peutherer 1989, p. 2). The mechanism that can secure this improvement is perceived to be a corporate GIS. As they go on to say, 'a corporate view is essential if the synergy is to be realised which will assist both the management of change and increase operational efficiency' (p. 13). Similarly Bromley and Coulson's (1989) study of the use of spatial data sets within Swansea District Council highlights considerable duplication which, it is argued, would be eradicated by a corporate GIS. This in turn is perceived to demonstrate the 'impracticability of introducing an isolated departmental GIS' (Coulson and Bromley 1990, p. 215).

The corporate approach is the epitome of the managerial rationalist perspective applied to GIS implementation. In many ways it is hard to refute the logic of the argument. If exchanging spatial data is the life blood of an organization, to inhibit the flow of such information must reduce organizational effectiveness. Failure to adopt a corporate strategy when the technology is capable of so doing might therefore appear plain stupid. However, it is important to consider whether the underlying assumptions presented in Figure 3.1 can be substantiated in practice. It is such questioning that provides the starting point for the social interactionist perspective on technological implementation.

Social interactionism

The underlying logic behind both technological determinism and managerial rationalism is based on a conception of how organizations ought to operate. In contrast, social interactionism starts by attempting to understand how the world is. Empirical studies have shown that it is frequently difficult to identify direct cause and effect relationships with respect to the implementation of computer-based systems. Detailed investigations have noted confusing and often contradictory experiences. For instance, studies that have examined the capacity of information technology to transform work practices have found that computers both centralize and decentralize authority, enrich and routinize jobs and create and destroy employment; sometimes simultaneously within the same organization. (For summaries of these investigations, see Kling 1980; Markus and Robey 1988; Rule and Attewell 1991.) These seeming contradictions highlight a number of important weaknesses in the technologically and rationally based perspectives on implementation.

The first issue concerns the extent to which the theoretical potential of a particular innovation will inevitably be realized in practice. Studies of the implementation of computers in a wide variety of organizational settings indicate that the outcome of this process is rarely entirely positive or for that matter completely negative (Kearney 1990; Lyytinen and Hirschheim 1987; Moore 1993; Mowshowitz 1976). It appears that the introduction of computer-based systems is much more a case of relative trade-offs between costs and benefits. Second, as the outcomes of implementing the same technology vary so markedly it would seem that it is not the innovation itself that determines the results of this process but rather the particular organizational and institutional circumstances. Even where computerization is part of a broader strategy of change within an organization it has been found that thorough pre-planning based on a rational management approach is insufficient to ensure the desired results in practice. This suggests that organizational rather than technological matters have a significant influence on the experiences of users. Long (1987) has gone so far as to conclude that office automation failures are only 10 per cent due to technical problems and 90 per cent to organizational and managerial issues. (See also Buchanan 1994; Eason 1993; Miles 1990.) As a result, while the introduction of new technology and the accompanying implementation strategy may act as instruments of change within an organization they do not of themselves cause that change or even in the vast majority of cases represent the catalyst of change (Danziger *et al.* 1982; Dunlop and Kling 1991b and 1991c; Kraemer 1991; Kraemer and King 1986; Rule and Attewell 1991; Scott 1990). As King states, 'Local government will shape technology to their agendas; not the other way round' (King 1985, p. 29).

The third weakness which has been implicit in much of what has been said above is a tendency for technological determinism and managerial rationalism to divorce technology from the jigsaw of conflict and cooperation within which they are expected to contribute (March and Sproull 1990). In both cases, not only are perceptions of technology narrowly focused on the equipment and methods but a similar mechanistic view is taken of the context into which such systems are expected to operate. This has profound implications for the style of implementation that is proposed as the propensity for individuals within an organization to defy the sentiments of a rationally based strategy are hardly considered. However, as Barrett and McMahon note in their studies in the health service:

> shifts in policy or demands for services are not spotted solely from the windows of the chief executive's suite of offices! The whole organisation is full of strategists, peering through windows at all levels of the organisation, all scanning, anticipating, planning and adapting: all seeing how primary changes will either be disadvantageous or offer opportunities for advancement. So the true impact of change 'out there' is not felt at some imaginary organisational boundary but instead it is felt 'in here' in how it affects the interests, power relationships and bargaining tactics of the organisation's partisans.
>
> (Barrett and McMahon 1990, p. 262)

This description of organizations suggests that the implementation of computer-based systems is far more complex and problematic than the preceding perspectives have implied. Individuals seldom respond to change in the manner laid down in formal strategies. Each member of staff within an organization assesses the personal implications of the introduction of a GIS, rather than the extent to which it is likely to enhance overall organizational performance. As a result, change, in whatever form, tends to be treated with suspicion (Keen 1981; Mumford and Pettigrew 1975; Robey 1987). Such a response should not be regarded as blind resistance to a new approach. In most cases scepticism or anxiety about a particular innovation reflects fundamental weaknesses or a failure to appreciate fully the social and political implications of the change. Keen (1981) has investigated the reactions of users, noting in particular their propensity to formulate counter-implementation strategies. These include such activities as the following: lying low so as to avoid direct confrontation while at the same time providing little encouragement; relying on inertia by prolonging the initial discussions to such an extent that everyone loses interest in the project; undermining the key individuals in the project so that they lose credibility; exploiting the system designers' lack of knowledge about the organizational environment and therefore ensuring that the eventual product is inadequate and has to be abandoned; or defining such ambiguous or alternatively ambitious goals that system development becomes virtually impossible. As a result, while there may be one formal strategy to guide implementation it is inevitable that numerous informal strategies will be simultaneously acted upon, each demonstrating an underlying concern about personal status, promotion prospects, job content or authority.

Professional associations also have a tendency to form strong communities of interest and are keen to articulate their concerns. The espoused organizational theory is therefore accompanied by a wide range of theories in use (Argyris and Schon 1978). Pfeffer (1978) even argues that the formal strategy is merely a rationalization of the controlling group's underlying objectives of reinforcing their existing position within the organization. Pinch and Bijker (1987) have similarly noted in relation to diffusion as a whole that perceptions of the value of a particular innovation are likely to vary greatly even between individuals within the same organization and therefore to conceptualize this process as linear in nature is an oversimplification. Historical evaluations of technological

Perspectives on implementation

diffusion indicate the multi-directional character of the process. In many cases the widespread adoption and implementation of an innovation is dependent not so much on the quality of the technology itself but rather on social attitudes or institutional developments (Bijker *et al.* 1987). As a result of such evaluations, Eason has stated:

> The description of the design process as a political process with implementation and counter-implementation forces at work will accord with the experience of many people engaged in the process, if not with the textbooks on system design, which rarely mention these realities of organisational life. It is clearly dysfunctional for the organisation to proceed in this way.
>
> (Eason 1988, p. 33)

It is this understanding of the sociopolitical realities of organizational life that provides the starting point for the social interactionist perspective on implementation. Table 3.4 provides a summary of the main characteristics of the approach.

The fundamental assumption of the social interactionist approach is that technologies are not independent of the environments in which they are located but rather only gain meaning from their context. The adoption and effective implementation of an innovation is therefore the result of interaction between the technology and potential users within a particular cultural and organizational arena (see, for example, Bijker *et al.* 1987; Dunlop and Kling 1991a; Goodman *et al.* 1990; Hirschheim 1985; Hirschheim, Klein and Newman 1987; Innes and Simpson 1993; Klein and Hirschheim 1989; Markus 1984). Implicit within these comments is the notion of technology as combining machines, methods and, most importantly, knowledge. Knowledge is not regarded as an independent variable but reflects an individual's interpretation of the norms and values of a particular social and/or professional grouping within their organizational context. As a result, Sproull and Goodman suggest that 'technology is a socially constructed reality' (Sproull and Goodman 1990, p. 259). Constant (1987) goes

Table 3.4 Main characteristics of social interactionism

Characteristic	Social interactionism
Propensity for adoption	Uncertain, depends on trends in society
Reason for adoption	To enhance symbolic status or power
Style of implementation	An organizational process that is problematic and uncertain
Constraints on implementation	Social and political processes
Likely outcome of implementation	Uncertain, at best a mixture of positive and negative results
Perception of technology	Machine, methods and knowledge
Perception of organizations	Cultures

further, arguing that potential users purchase an image of the technology, particularly the output of that system, rather than actual items of machinery. Very often therefore, for a technology to become widely adopted it is more important that it becomes fashionable than that objective measures of its utility indicate a considerable advance over existing techniques. The symbolic value of the technology is therefore crucial to adoption and utilization. One of the greatest achievements of the computing industry has been to link such technologies with 'the symbolic politics of modernism and rationality' in such a way that they 'offer the giddy excitement of adventure with (at the same time) the liberating lure of new possibilities that have few associated problems' (Dunlop and Kling 1991b, p. 7). The real power of this imagery lies in the extent to which it convinces those in control that their position will become increasingly untenable if they do not embrace such technologies (Pfeffer 1978 and 1981). However, acquisition does not necessarily imply that the controlling coalition or staff in general have also embraced the changes that are associated with achieving utilization. Moreover it is likely that there will be a wide variety of underlying perceptions and expectations of a particular technology even within the same organization.

The view of innovation as socially constructed reality suggests that technologies are not value-neutral. The introduction of new computer technology is therefore loaded with social and political meaning. For the social interactionist perspective, therefore, questions of ethics and values are not external to the technology but inherent within the system (Ladd 1991; Mouritsen and Bjørn-Andersen 1991). This is mirrored in the work of Rogers (1983) which suggests that the key characteristics of technologies which have been widely diffused are essentially social rather than technological. These include simplicity, observable benefits, relative advantage, ability to make small trials and compatibility with existing social norms. Once again the propensity for effective implementation to occur appears to depend on the unique sociopolitical characteristics of each organization. Given the huge differences between contexts and the widely varying values and motivations of the individuals within these environments, the processes of adoption and implementation must be both complex and problematic. No universal claims are made about the likelihood of achieving a favourable result or whether or not the outcome will be beneficial. In contrast the expectation in any one environment is that there will be both winners and losers, with the ultimate balance determined by the specific circumstances. Consequently, even in relation to exactly the same technology the response of potential users is likely to vary considerably between, as well as within, organizations.

The conception of organizations as unique cultures is implicit in much of the discussion concerning social interactionism. This in turn has significant implications for implementation as such an understanding of the nature of organizations renders universably applicable strategies at once impractical and in some senses foolhardy. Panaceas do not exist and, as Handy argues of management in general, 'many of the ills of organizations stem from imposing an inappropriate structure on a particular culture, or from expecting a particular culture to thrive in an inappropriate climate' (Handy 1993, pp. 180–1). The management of change, which in effect underlies implementation, must be nurtured and cajoled rather than imposed and controlled. Furthermore, given the lengthy time span involved in securing routine use, it is likely that an approach to implementation that is out of tune with its organizational environment will fail at one of the many hurdles that will undoubtedly be encountered. Understanding the organizational context into which a technology is to be embedded is therefore regarded by the social interactionist perspective as crucial to system development. One approach, which has gained considerable attention particularly in the GIS field, is based on the identification of a 'champion' or 'change agent' (Beath 1991; Kanter 1983). The rationale behind this is that adoption and utilization are dependent upon the presence of a charismatic individual. Such individuals are usually characterized as a senior manager, project leader or a lay politician and as the sort of person who relishes a battle and usually has the tactical awareness to realize their goals. Some might argue that their objectives have more to do with personal ambition rather than the well-being of the organization. Nevertheless the idea of champions has tended to become a cookbook remedy for dealing with the organizational aspects of technological implementation. In many situations an individual can be identified who has played a disproportionately significant role in a particular project. However, this is rarely sufficient on its own to guarantee success (Eason 1994). Moreover, such an approach will only yield the desired result if it is in accord with the organizational culture. It is this crucial final point that dominates the social interactionist approach to implementation.

The underlying goal of a social interactionist style of implementation is to gain organizational and user acceptance for the technology, not simply to achieve a technically operational system. Furthermore, it is assumed that routine use will not take place unless the system meets the fundamental needs and capabilities of users and has obtained their commitment and support. As a result, emphasis is placed on participative approaches to implementation typified by user-centred design philosophies (Eason 1988) which regard the introduction of computer-based systems as part of a process of organizational change (Hirschheim 1985; Pettigrew 1985 and 1988a). The whole process of technological innovation is therefore viewed not in isolation from the organization but as part of a process of strategic change: a process that, if it is to yield benefit, must acquire legitimacy.

A user-centred design philosophy starts from the assumption that it is organizational rather than technological issues that are most likely to threaten the effective implementation of computer-based systems. This philosophy does not deny that there is a technical component to implementation but argues that to place the focus of activity on this element is liable to be counter-productive. Moreover, it is crucial that users feel that they have ultimate ownership of the project if they are to employ the system in their daily activities. User involvement in this case amounts to much more than token consultation and as a result these individuals need to be given sufficient time and skills that they have real ownership of the project. In these circumstances it is important that users participate fully in the decisions that precede the acquisition of the technology as well as the subsequent phases of development and evaluation. This implies that organizational learning must proceed alongside technical design and not behind it, as the earlier perspectives have suggested. Such an approach to implementation is also likely to result in crucial issues such as the impact of the technology on job content, lines of responsibility, status and remuneration being considered at an early stage. It is also probable that the provision of appropriate forms of training will be given a high priority. Participation in training courses frequently represents the first interaction that users have with the technology. If they fail to be relevant or are targeted at the wrong level, such courses will alienate the people who are actually expected to use the system. As a result, on-the-job training undertaken by a familiar and popular member of staff, although not necessarily the most technically gifted, is often the most effective approach.

Given the complexities of organizational life, special priority must be given to developing commitment to a project through team building and the formulation of informal plans rather than devoting the resources of a few to building all-encompassing strategies which are likely to be quickly superseded by events. At the same time it is appreciated that few organizations can tolerate anarchy, therefore more informal forms of strategy building must be set within a widely shared vision of the overall objectives of the organization. Vision in this sense is inseparable from the underlying norms and values of the organizational culture which is not necessarily reflected in statements made in publicly available documents.

Eason (1988 and 1994) identifies five distinctive strategies in terms of the overall style of implementation. These range from the 'big bang' approach, through less traumatic methods such as running manual and computerized systems in parallel or the phased introduction of the new technology, to more evolutionary approaches which include periods of trials and dissemination or an entirely incremental structure. The more evolutionary and incremental approaches tend to be favoured as they provide greater scope for users to learn and adjust to the proposed development. Moreover, it is through these mechanisms that a project can gradually gain irreversible momentum. It is implicit that the process is as important, perhaps more so, than the final product. The change implied by the introduction of information technology is therefore viewed as embedded within the organizational politics of the particular environment (Pettigrew 1988a). This final point is crucial for, while incremental approaches are considered as the most likely to secure user acceptance of the technology, it is simultaneously acknowledged that such strategies will only succeed if they are in tune with the organizational culture. For this reason the introduction of technological innovations may equally be alien to some organizations.

The recommendations of the social interactionist perspective are tempered by an overarching analytical framework which acknowledges that what is possible as well as appropriate depends on the particular organizational circumstances. This often leads to criticism that while such perspectives provide an explanatory

Perspectives on implementation

framework they lack precise prescriptive advice and are therefore of only limited value. A user-centred philosophy, for example, offers no cookbook recipes but neither does it assume how the world ought to be. In the end the appropriate approach is regarded as largely dependent on the capabilities and needs of users and the resources that are available in the organization in question. As a result it is assumed that without user acceptance of the innovation it will be difficult to achieve utilization. The next section will explore how this analytical framework has been applied to GIS.

Implications for GIS

The most notable feature of the social interactionist perspective in relation to GIS is its virtual absence from the literature in this field, either in terms of an explanatory framework or as a basis for implementation. The preceding discussion has noted that the majority of the studies related to GIS have concentrated on refining technical know-how, largely divorced from the environments in which the technology is expected to operate. Moreover those that have considered how best to tackle the introduction of such systems have assumed that rationally based management strategies will be able to realize the desired result. This largely technological orientation would seem to be partly symptomatic of the relative newness of GIS as a commercial product as well as the lack of exposure of those concerned with the findings of studies conducted in relation to information technology in general. Some might argue that there has been a tendency to turn a blind eye to the realities of implementing GIS technologies in the vain hope that they will disappear. However, with the experiences of users increasingly emphasizing the difficulties of implementing such systems and the uncertainty of securing the expected potential, the underlying assumptions are increasingly being questioned (Budic 1994; Campbell 1990a, 1991 and 1992b; Crosswell 1991; Lopez and John 1993; Masser and Onsrud 1993; Medyckyj-Scott and Hearnshaw 1994; Onsrud and Pinto 1991; Onsrud and Rushton 1995; Openshaw, Cross, Charlton, Brunsdon and Lillie 1990; Peuquet and Bacastow 1991; Worrall 1994).

Figure 3.2 points to some of the key weaknesses in the accepted wisdom based on a corporate implementation strategy. The first concerns the basic practicalities of corporate working and, by implication, the challenge that this poses to existing practices and relationships. For instance, it is relatively easy to devise a strategy for spatial data standards on paper but it is likely to be considerably more difficult to secure widespread organizational acceptance, no matter how logical it may appear to its creators. The essence of the corporate approach is that the introduction of GIS technologies should be used as a catalyst to change the manner in which activities ranging from daily operational tasks through to strategic decision-making are undertaken. However, as the earlier discussion has emphasized, change is a matter of organizational politics, not managerial rationality. If the proposed change is out of step with the norms and

GIS and organizations

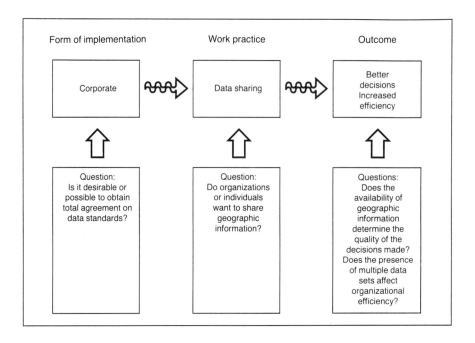

Figure 3.2 Limitations in the underlying rationale of the corporate approach to GIS implementation

values of the organizational culture it is likely that those held responsible will encounter numerous counter-implementation strategies. For instance, it is possible within most organizations to identify distinct professional groupings such as accountants, salespeople or personnel specialists who tend to assert themselves through departmental rivalries. These subcultures each have their own priorities, skills, language and even rituals; characteristics that they are often anxious to preserve. In these circumstances, while the introduction of a corporate GIS would not be impossible, its realization would undoubtedly take considerable awareness of and sensitivity towards the sociopolitical processes inherent within the various subcultures.

Fundamental to the logic of adopting a corporate approach to GIS implementation is the underlying assumption that only a corporate GIS can facilitate data sharing and, perhaps more profoundly, that procedures to enhance the exchange of geographic information are highly valued by individuals within organizations. Information is therefore viewed as a corporate resource. At a practical level, the data-sharing arguments are based on an assumption that the use of computers to store and transmit information is likely to be more effective than filing cabinets, hard copy memorandums and word of mouth. However, this can only be the case if the necessary technology is physically accessible and easily usable by non-technical specialists. Moreover, for technology to provide a real advance the type of information that is stored in such systems must be able to provide the complete answer to an individual's enquiry. If, for example, the answer is only partial it would be more efficient for the individual concerned to have consulted their traditional source. It is also worth noting that if a wider perspective of the daily experience of work of most people is examined, it is often only the contact with other staff in the organization that makes the tedium tolerable. This emphasizes again the need to consider the impact that technology will have on work practices and, more importantly, on job content.

The assumptions concerning GIS and data sharing not only have practical implications but, perhaps more fundamentally, economic and political ones. Masser and Campbell (1994) have pointed out that in economic terms there is a tendency to treat geographic information as if it were a public good, that is to say something that is accessible to everybody and which can be shared without loss to any individual. Yet even within the same organization information often takes on the characteristics of a commodity (Bates 1988; Openshaw and Goddard 1987). For instance, the willingness of individuals to allow their information to be placed on a corporate database is often dependent on their personal perception of the costs and benefits. In many cases the benefits of data sharing are not realized by the individuals who bear the costs, particularly the time costs of inputting the data. Consequently, if the direct benefit to the individual is limited they may be unwilling to participate, therefore diminishing the whole. A less confrontational approach would be to input the data but be unconcerned about such matters as data accuracy.

The final set of assumptions concerning data sharing centres on the political properties of information. As March and Sproull state, 'Information in organisations is not innocent; rather it is shaped by expectations of its consequences' (March and Sproull 1990, p. 151). It has been argued that existing flows of information have evolved in response to the balance of power (Mosco and Wasko 1988). Any changes can therefore either be interpreted as an attempt by the dominant coalition to reinforce their position, thereby increasing the divide between the information-rich and the information-poor (Pickles 1991 and 1995), or alternatively as a challenge to the *status quo* (see, for example, Curry 1995; Lake 1993). Whichever interpretation is preferred, proposals to integrate formally isolated data sets have significant implications for the ownership and control of information and consequently the distribution of power. Moreover, the theoretical advantages of sharing information may be viewed less positively if individuals feel that it will lead to closer supervision of their activities or open up the decision-making process to greater scrutiny. In such circumstances of fear and suspicion the chances of counter-implementation strategies being put into effect look probable rather than merely possible. The logic of data sharing as an enhancement of work practices is therefore likely to confront practical, economic and political hurdles when it encounters the realities of organizational life.

It is not surprising, given the weaknesses that have been identified in the implementation strategy and proposed work practices associated with a corporate GIS, that questions should also be posed of the assumptions concerning the outcomes of this process. Implicit within the argument that greater availability of geographic information results from a corporate GIS is the notion that spatial data determine the quality of decisions made. However, the discussion in the previous chapter has demonstrated that information is seldom the critical factor in deciding upon the most appropriate course of action. Decision-making has been shown to be a far more complex and problematic process than suggested by the oft-heard dictum 'better information leads to better decisions'. Even at the most basic level it is virtually impossible to discern what are better decisions, while data overload is frequently more of a problem to decision-makers than scarcity of information. Furthermore, information cannot simply be thrown at a problem in order to produce solutions. Potential users need information to be in an appropriate form, available at the right time, in the right place and, most importantly, to support their case. GIS may be said to unlock vast reserves of spatial data, but for what purpose? One of the most profound problems for those involved in designing decision support systems is how to anticipate the issues for which such systems could make a contribution and therefore be able to respond within the time frame of the decision-maker. It is even possible that reliance on computing technology may, by giving decision-making a more technical orientation, tend to thwart creative thinking (Postman 1992; Roszak 1994).

The second area in which a corporate GIS has been identified as a significant advance is in combating duplication of effort. This argument is usually based on surveys that have found multiple copies of the same data set within a particular organization (see, for example, Bromley and Selman 1992; Coulson and Bromley 1990). It would in many ways be surprising if this were not the case and there is a danger in assuming that efficiency can be enhanced by making this data available through a common medium. For instance, while several individuals may refer to the same data set there may be profound differences in their actual needs, such as the level of accuracy and up-to-dateness required, the exact part of a data set utilized and the manner in which it is employed. As a result, while staff in many sections of an organization may be utilizing the same data set, in practice each is using it in a highly personalized manner. Given much of the preceding discussion about organizational cultures it is worth bearing in mind the following comments in relation to efficiency:

> New technologies that alter attention patterns disrupt existing organisational systems. Such disruptions in procedures, role relationships, hierarchies, incentives and memories are not neutral with respect to their impacts on the social order of an organisation. It is rare the executive who embraces revolution in order to increase efficiency.
>
> (March and Sproull 1990, p. 163)

These arguments are summarized in Table 3.5. It highlights the extent to which a corporate approach may actually fail to deliver the expected outcomes. Moreover, in contrast to the traditional wisdom on GIS implementation the

Perspectives on implementation

Table 3.5 Arguments that question the appropriateness of a corporate strategy towards GIS implementation

Disadvantages of a corporate strategy	Advantages of separate system development (a departmental approach)
Variations in priorities among users	Increased independence
Disagreements over:	Sensitivity to user needs
• data standards	Control over priorities
• access to information	Clear lines of responsibility
• leadership	
• equipment	
• training	
• job content	
Economic costs of sharing spatial data	
Challenge to the existing balance of power	
No improvement in decision-making	
Decreased organizational efficiency	

introduction of separate systems, particularly at a departmental level, may offer some advantages in terms of the capacity to achieve utilization. It is important not to underestimate the significance of issues such as independence and the potential for greater sensitivity to user needs and capabilities which are inherent in such an approach. The implementation of systems in this context also removes the sense of mistrust that is often associated with what is seen as imposition from above. As a result it is likely that the propensity for counter-implementation strategies to evolve will be lessened. Given the limited research that has been undertaken it is not possible to substantiate these comments. However, it seems probable that the extent to which such an essentially fragmented style of implementation can be tolerated will depend on the nature of the organizational culture. It is also possible that while the number of variables involved are severely reduced by a departmental style of operation, such environments may in microcosm exhibit the same interaction of social and political processes that have been noted at the larger scale (Campbell 1990a).

Despite this critique of the managerial rationalist style of GIS implementation there have been few attempts to devise alternatives that take the realities of organizational life as the starting point. Work by the authors has, however, identified three areas that need to be considered if effective utilization is to be achieved (Campbell 1990b; Masser 1992). These are as follows:

1. an information management strategy that identifies the needs of users and takes account of the resources and values of the organization;
2. commitment to and participation in the implementation of the system by

individuals at all levels of the organization; and
3. an ability to cope with change.

This work suggests that while the precise mechanisms to tackle each of these areas will depend on the nature of the organizational culture, it is unlikely that implementation will be secured without detailed consideration of these issues. The following discussion will explore each in turn.

An information management strategy

Any information system will only be utilized if it generates information that is of value to users. It is therefore regarded as important that emphasis is placed on identifying the information priorities of the organization and how such resources should be incorporated into the work of staff. Once the information needs of the organization have been established it follows that decisions can be made about the most appropriate methods of storing and handling this information. Moreover, it is important that such judgements are based on a realistic assessment of the available resources including skills, personnel, experience, equipment, data, training facilities and funding as well as the values and practices of the organization. Consideration of the technology therefore follows from the identification of the information priorities of the organization and the available resources and culture. It is also important to assess how this information is used, in what form, why it is needed and by whom as well as the interests that would be affected by a change in work practices. For instance, does the type of use translate into a computer application or simply a requirement for a filing cabinet? It is quite possible, therefore, that while geographic information is widely used within a particular organization, a computer-based system such as a GIS is seen as offering few benefits and perhaps considerable disruption. It is implicit, given the nature of the issues that must be considered, that the central involvement of users and senior management is crucial. They must feel that they have ownership of such a strategy and share in the sentiments that are expressed. In a sense, therefore, it represents the embodiment of the norms, traditions and values of the particular organization. The chances of systems becoming redundant would seem to be high without a widely accepted sense of the information needs of both the organization as a whole and individual members of staff.

The use of the term 'information management strategy' implies a comprehensive written document; however, the scope and form of the knowledge that has been identified as constituting such a strategy may take a wide variety of forms. At its most comprehensive the strategy would indicate the information priorities of the whole organization, appropriate methods for handling these data and the necessary training provision, with each department identifying their particular needs. In other cases a more fragmented pattern may develop where each department and perhaps even subsection of a department develops their own separate strategy. A further issue is the status of these strategies, that is to say to what extent they have been formalized within the procedures of the organization. For instance, has the strategy received full approval or is it regarded as a working document? In some cases, formalizing the strategy may be perceived as drawing too much attention to the task and as such may be likely to lead to a virtually meaningless outcome, while in other circumstances it might be deemed fruitful to, metaphorically, 'bang heads together'. User departments may be the most reticent about formalizing their strategies as they are likely to conflict with central thinking, particularly that of the computer specialists. The final issue is the form of these strategies. The term 'strategy' implies a written document but it is quite possible that an information management strategy may take a less tangible form. For instance, priorities and attitudes towards information may be deeply embedded within the culture of the organization, with new staff automatically socialized in such traditions. Similarly, it is possible that an individual is perceived to be the embodiment of the organization's information management strategy with their verbally expressed views holding sway. The critical issue would seem to be that a robust framework has been established which is widely shared by staff at all levels of the organization rather than the existence of a highly detailed and lengthy document. Overall the process is far more important than the final product. The issues involved concern the fabric of the organization, not simply matters initiated in response to the introduction of GIS technologies.

Commitment and participation

The second area has already been raised in relation to the information management strategy, that being the need to ensure the commitment and participation of users. This aspect of the process very much reflects the issues raised by the user-centred philosophies with respect to the implementation of computer-based systems in general. User acceptance and commitment to GIS are unlikely to be achieved if they feel excluded from the decision-making processes that will shape their daily work experience. As a result it is important that mechanisms are devised that facilitate widespread participation. In the case of GIS such an approach appears crucial even in purely practical terms as the length of time required to develop an operational application as well as the ongoing costs of maintenance suggest that commitment will need to be sustained over a considerable period of time. In such circumstances support for the GIS will need to be able to withstand significant periods when scarce staff time and financial resources are devoted to system development, yet little appears to be delivered in return. The precise nature of the mechanisms adopted to facilitate participation will depend on the organizational culture as was outlined earlier. However, it is important to recognize that GIS are often seen as a multi-purpose technology which can satisfy the needs of a variety of users. As a result, commitment and participation will need to be achieved both vertically in terms of a variety of skills and horizontally among what is likely to be a heterogeneous set of potential users.

Change and instability

The final area of concern focuses on the issue of organizational change. Although this has been implicit in the previous two chapters its significance is such that it deserves separate attention. Change and uncertainty is an inherent part of organizational life. Moreover it is clear that the introduction of a computer-based system such as a GIS contributes to the disruption that is already faced by the organization. In these circumstances it is also inevitable that during the lifetime of the GIS, perhaps even the initial period of implementation, the goals and roles of the system will change. The implementation and long-term utilization of a GIS is therefore dependent on the ability of the organization to cope with change. An important aspect of the art of implementation is an appreciation of the level of change and disruption that a particular organization can tolerate before it becomes viewed as a threat and the whole programme is aborted. This in turn implies that the types of application that can be sustained in one organization may be impossible in another. As a result, if GIS are to be utilized in some contexts, it will be important that the change in work practices that is associated with implementation is kept to a minimum.

Summary

The terms used could be found in a managerial rationalist cookbook but, as the above discussion indicates, it is the manner in which they are applied that is crucial. An information management strategy in this case is not part of a highly structured design method that assumes the needs of users; rather it may be fragmented and informal. Similarly, achieving commitment and participation is seen as profoundly important, involving the empowerment of users and not mere token gestures. As a result the issues raised point to the importance of starting with the realities of the organizational world rather than how it should operate in theory. The growing list of high-profile computing disasters suggests that failure to recognize adoption and implementation as organizational processes is to court failure. This does not imply that to establish the information priorities of the organization, ensure user commitment and participation and be sensitive to the issue of change will automatically guarantee utilization, but that to fail to consider these issues is likely to threaten the integrity of the project. In the end the social interactionist perspective suggests implementation to be social and political in nature, not rational and technical. As Hirschheim states, 'Information systems are not technical systems which have behavioural and social consequences, but are social systems which rely to an increasing extent on information technology for their function' (Hirschheim 1985, p. 278).

Success and failure

One issue that has been taken for granted in the preceding exploration of various perspectives on implementation is the notion of success and failure. There is a great deal of ambiguity in the use of these terms in relation to information technology. There are likely to be considerable differences in opinion even with respect to the same system, as the introduction of any technology brings with it both benefits and problems and therefore winners and losers. Those involved are also likely to have very different perspectives according to their area of interest. For instance, a computer specialist may be delighted with the elegant algorithms that have been devised to underpin a system and consequently may view it as a success, while someone who loses their job as a result of them is likely to regard the same development as a disaster.

In their review of the literature Lyytinen and Hirschheim (1987) identified four categories of failure:

1. correspondence failure;
2. process failure;
3. interaction failure; and
4. expectation failure.

Correspondence failure is concerned with the extent to which a system meets the objectives specified at the outset of the project. Such an approach very much reflects the rationality inherent within much management science, as it assumes that clear, unambiguous goals were set at the start of the project and that the environment can be controlled to such an extent that they remain relevant. The second category of failure refers to a situation where an operational system cannot be produced within the initial budget guidelines. The overspend is therefore viewed as a failure in the implementation process, more particularly the predictive capacities of management science. Again, there is a strong management science orientation to the underlying evaluation in that there is an inherent confidence in techniques such as cost–benefit analysis. Moreover, to assess success and failure in purely monetary terms ignores less tangible benefits. The third category, termed 'interaction failure', focuses on the extent to which a new system is used. Use in this case tends to be associated with user satisfaction and increased organizational performance. However, this could be misleading if there is no alternative but to utilize the technology. The fourth category is regarded as embracing the previous three and is based on whether the system fulfils expectations. The key issues in this case are deciding whose perceptions should be paramount and judging the relative differences between the expectations of the individuals concerned.

This categorization of criteria demonstrates the complexity of applying the terms 'success' and 'failure' to the implementation of information technologies such as GIS. It would therefore seem to be important that the boundaries of the analysis are kept as wide as possible as the impacts of computer-based systems often extend far beyond their particular focus of action. Consequently a GIS could be regarded as successful in its own terms yet may have caused massive organizational disruption. It is also worth bearing in mind that, given the often symbolic function of many computer-based systems, it is quite possible that the publicly expressed objectives and expectations may have little relationship with the underlying motivations. The purchase of GIS could therefore reflect a statement of modernity rather than any real intention as to how the system is to be used.

This analysis suggests that terms such as 'success' or 'failure' should be used with extreme caution. The limitations of each of the categorizations outlined above also suggests that insight is likely to be enhanced by maximizing the range of dimensions that are examined.

Conclusion

The three perspectives of technological determinism, managerial rationalism and social interactionism highlight the variety of interpretations that have been developed of the relationship between technology and organizations and also the consequences that this has for the implementation of innovations. The review of the perspectives has sought to highlight the various assumptions that underpin each approach. This in turn provides an explanatory framework for the subsequent investigation which seeks to describe and understand the experiences of those attempting to introduce GIS technologies into their organizational culture. This framework is crucial, for without a proper assessment of the underlying assumptions there is a tendency for the discussion to deteriorate into the type of debate outlined by March and Sproull, whereby 'Enthusiasts for a new technology are likely to attribute slowness in adoption either to correctable faults of the technology or to various forms of irrational resistance. Skeptics are likely to attribute failures to the fact that the technology either addresses a real need poorly or solves a problem that does not exist' (March and Sproull 1990, p. 147). The critical question for GIS is to understand what is actually happening in practice.

CHAPTER 4

Organizations and GIS: a case study of British local government

Introduction

Everyday references to diffusion evoke images of a race in which the objective is to pass some unspecified threshold of utilization indicating society's acceptance and, perhaps more importantly, dependence upon that innovation. However, the history of technological innovation is littered with a mixture of false starts, promising beginnings, unrealized expectations and fast finishes despite considerable doubt on the part of critics. In most cases the race itself appears to take the form of a marathon run over hurdles rather than a straightforward sprint. The question for this study is what form is the GIS race taking, and more particularly, what processes are affecting its course and outcome within individual organizations. The preceding discussion has highlighted a range of perspectives on the nature of implementation. Many questions have been raised. For example, is GIS utilization dependent upon the technological characteristics of the innovation, rational management strategies or a less predictable combination of social and political processes? How are users within particular organizational settings responding to the potential of GIS technologies? To what extent are the facilities offered being reinvented in each context? In the final analysis, will users assess GIS to be an innovation or irrelevance?

This chapter provides a link between the theoretical discussion and the findings of the empirical investigations which have explored the relationships between innovations and organizations. The nature of GIS as an innovation was considered in Chapter 2; it is now important to turn to the organizational context that forms the focus for the research, namely British local government. By way of background the key features of local government in Great Britain will be outlined with particular attention focused on the style of internal management and the extent to which there are similarities with other groupings of

GIS and organizations

organizations. The final part of the chapter highlights the main features of the research methodology.

British local government

Local authorities as organizations

The objective of this review of British local government is twofold: first, to outline the key features of this sector in order to clarify the terminology that will be used subsequently, and second, to explore the nature of local authorities as organizations and therefore the likely generalizability of the findings. Consideration of any grouping of organizations often gives rise to the seeming contradiction of simultaneously outlining the common framework within which activities are set, yet also emphasizing the uniqueness of each environment. For instance, all British local authorities operate within the same basic external framework yet overall there is a huge diversity of approaches. Similarly, such richness and variety typify the companies and agencies that make up sectors such as car manufacturing, food retailing, telecommunications or care of the elderly. Moreover, often a company in one sector may have more in common in organizational terms with a concern operating in a different area than those involved with similar activities. As a result the characteristics of the external framework are less important than the processes that affect how these constraints and opportunities are internalized within a given organization. Like any other grouping of organizations, local authorities consist of people performing tasks. It is the manner in which these activities are undertaken that varies greatly.

Structure of British local government

The structure of local government in the United Kingdom takes three distinct forms. Scotland and Northern Ireland each have their own styles of operation, while England and Wales share a common structure. As a result of the special circumstances in Northern Ireland the functions of local government in this case have largely been taken under the control of central government. It was not therefore appropriate to include Northern Ireland within the study.

The system with which the majority of the population in Great Britain are most familiar is that found in England and Wales. Even in this case a dual system operates whereby metropolitan and non-metropolitan areas have separate structures. For the metropolitan areas, which comprise the six large conurbations and London, there are 36 metropolitan districts and 33 London boroughs. These are all subsequently referred to as metropolitan districts (see Figure 4.1). Since the abolition of the upper-tier metropolitan counties in 1986, these multi-purpose unitary authorities have worked alongside a range of *ad hoc* bodies which were introduced at the time to coordinate metropolitan-wide activities such as

A case study of British local government

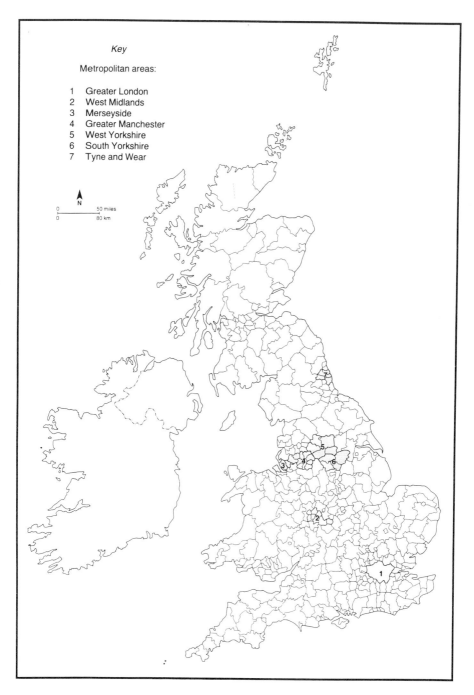

Figure 4.1 District authorities in Great Britain in 1994

GIS and organizations

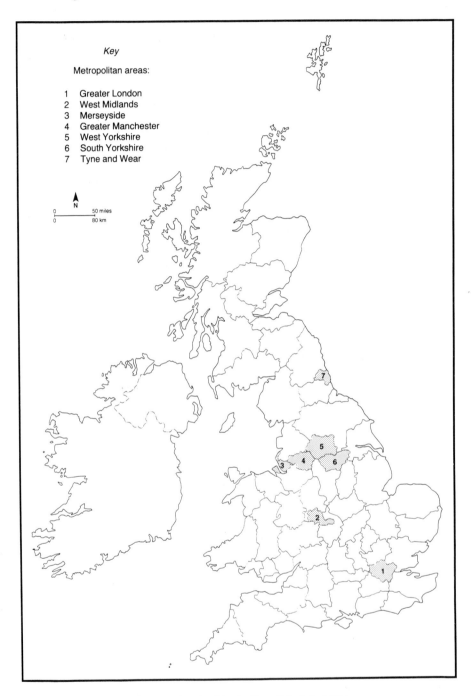

Figure 4.2 County and regional authorities in Great Britain in 1994

passenger transport and the police and fire services. In contrast, in the non-metropolitan areas there is a two-tier structure consisting of 333 districts and 47 counties. These authorities will subsequently be referred to as shire districts and shire counties. The Scottish system, in contrast, makes no distinction between metropolitan and non-metropolitan areas, having a two-tier structure of 53 districts and 9 regions. There are, however, three unitary authorities for the island communities which depart from the general structure. Figures 4.1 and 4.2 illustrate respectively the lower-level authorities consisting of the metropolitan, shire and Scottish districts, and the higher-level authorities comprising the shire counties and Scottish regions.

The division of functions in the non-metropolitan areas results in districts largely taking responsibility for local-scale services such as housing, public health, refuse collection and local land use planning. In England and Wales the higher-level county authorities have responsibility for the provision of services such as education, roads and transport, strategic land use planning, the emergency services and social services, while in Scotland the regions have a rather broader role, including such activities as supplying water. As a result of this division of responsibilities and the difference in size, the counties and regions command much larger budgets than the shire and Scottish districts.

In contrast to many other contexts British local authorities have responsibility for a relatively limited range of activities (Batley and Stoker 1991). For example, provision of most health care facilities, energy-related services and, in England and Wales, water supply is undertaken by either unelected regional bodies or privatized companies. Furthermore, there is no tradition of maintaining a cadastre at the local level in Great Britain.

A second distinctive feature of the British local government system is the size of the basic unit of authorities (see Table 4.1). The average populations of the Scottish, shire and metropolitan districts at the time of the 1991 census were respectively 94,300, 96,700 and 253,600. Norton (1991), in his comparative study, notes that reorganization in West Germany and Scandinavia resulted in the

Table 4.1 Average population of local authorities in Great Britain in 1991

Type of local authority	No. of authorities	Average population in 1991
Shire districts	333	96,700
Metropolitan districts	69	253,600
Shire counties	47	685,000
Scottish districts	53	94,300
Scottish regions	9	547,400
Scottish Islands	3	23,900

Source: 1991 Census of Population

GIS and organizations

populations of the basic unit of local government rising to 8,000, 10,000, 12,000 or 20,000. It is clear that these authorities in no way approach the size of the much larger units that have evolved in Great Britain. For example, the smallest shire district of Teesdale had a population of 24,068 in 1991. In contrast the populations of counties and regions are smaller than the highest-level subnational government in many countries, with average populations in 1991 of 685,000 and 547,400 respectively. This reflects the absence of regional government in Great Britain. (For a more detailed comparative analysis of local government, see Batley and Stoker 1991.)

The appropriateness of the two-tier structure of local government in the shire counties and Scotland has recently become a cause of concern for the government. As a result in 1991 it initiated a full review of the current arrangements. Draft proposals for Scotland and Wales have already been published, although consultations are still taking place and the precise form of the final legislation is as yet unclear. The process in England and Wales is being undertaken on a county-by-county basis with the initial phase of devising proposals not yet complete. It appears likely that the outcome of this review will be a patchwork consisting of unitary authorities for the large urban areas and a mixture of single and two-tier local government for the remainder of the country. However, in organizational terms the most important element of this review is the uncertainty that has been engendered rather than the precise pattern that will result. Inevitably, this has led many authorities to concentrate on defending their actions in an attempt to ensure their own survival and in some cases to focus on initiatives that will yield results in the short term.

Internal management

Initiatives to alter the manner of internal management of local authorities have tended to accompany reviews of the structure of local government. This is the case at the present time with profound changes taking place which affect the whole nature of governance at the local level. It is, however, important not to overestimate the impact of these changes on local authorities as organizations. Change, and the uncertainty that is associated with it, have tended to be a relatively constant aspect of the environment in which authorities must operate. Moreover, such circumstances are common to virtually all organizational settings. The following discussion highlights the main trends in public sector management, although it is emphasized that this is not a complete evaluation of the theoretical implications of the developments that have taken place (see Leach, Stewart and Walsh 1994).

The first active interest in the organization of local authorities by central government came in the late 1960s and early 1970s alongside reform of the overall structure. Efficiency was the sentiment that underpinned the subsequent proposals. The result was fewer larger authorities but such structural changes were not regarded as likely to yield the desired benefits without the appropriate arrangements for service delivery. A number of studies were commissioned to inform the debate (see, for example, the Maud Report 1967; Bains Report for England and Wales 1972; and the Paterson Report for Scotland 1973). The subsequent reports were strongly influenced by contemporary management science thinking which emphasized the efficiency gains that would result from adopting a corporate approach. As a result, authorities were encouraged to appoint chief executives who, in conjunction with a small management board consisting of a few chief officers, would coordinate the delivery of services. This pattern was mirrored at the political level by proposals to establish central policy and resources committees which would take responsibility for identifying the council's priorities and thereby allocating resources. The overall aim was to devise a more centralized organizational structure through which an efficient and comprehensive approach to service delivery could be assured.

The managerial structures that evolved during the 1970s testify that the vast majority of local authorities attempted, at least in theory, to introduce more formalized and centralized patterns of working. Such reforms of organization have been regarded as echoing trends within the manufacturing sector, resulting in the efficient mass production of services. Moreover, the copious hierarchies and procedures that are associated with this approach have become inescapably linked with notions of bureaucracy. Hambleton and Hoggett (1987) refer to the relationship that developed between local authorities and their communities as 'bureaucratic paternalism'. However, bureaucratic forms of control based on the rational strategies inherent within the management science tradition proved difficult to realize in practice. Despite the presence of formal mechanisms for the centralized coordination of activities, service provision in most authorities was dominated by largely autonomous departments. Most departments centre on a specialist professional grouping such as surveyors, architects, social workers, land use planners or housing managers and, rather than all working towards a common goal as was envisaged, each focused on promoting their own interests. Hoggett (1991) views the tendency towards segmentalism not as a challenge to bureaucratic forms of organization but as an inevitable accompaniment to centralization. He draws on the work of Kanter (1983) in the private sector, arguing that functional specialization is an inherent part of complex bureaucracies and where this overlaps with professional groupings the result is likely to be a focus on occupational interests rather than those of the company or the local authority. At the level of the individual, loyalty to the profession is likely to be regarded more highly than loyalty to the organization. The programme of training involved in gaining membership to a particular profession often incorporates a subtle process of socialization. As a result, fellow professionals usually share a common educational experience, set of skills and, most importantly, language.

The nature of bureaucratic organizations provides them with few mechanisms with which to counter such challenges to their integrity. In most cases their main response is more of the same, that is to say tightening control through regulation and a greater array of procedural devices. Inherent within such actions are the seeds of their own destruction as professional interest groups inevitably try to side-step attempts to reinforce central control. As a result the use of the term 'bureaucracy', which is inextricably linked with local authorities, was increasingly deployed in a derogatory sense rather than Weber's notions of the only legitimate means of exercising authority. Certainly there was a tendency to dwell on matters of administrative detail rather than formulating policy initiatives of a more strategic nature. It is also apparent that, despite the subsequent pressures for change outlined below, it is this image of local authorities that persists among the vast majority of the population. Moreover, such perceptions are often unfairly associated exclusively with the public sector. (See, for instance, Deal and Kennedy 1982; Handy 1993; Kanter 1983 and 1990 on the private sector.)

During the 1980s bureaucratic forms of organizational control within the public sector came under attack from both the left and right in British politics. Some have associated the calls for more flexible organizational structures with the pressures created by an increasingly volatile world economy (Hoggett 1987). There is little doubt that management theorists like Peters and Waterman (1982), who were challenging the appropriateness of the management science approach within the private sector, also had a significant influence on public sector organizations such as local authorities. However, the extent to which the resulting patterns of working can be described as post-Fordist is open to question (see Burrows and Loader 1994). Much of the change in structure and organization that has taken place within local government is associated with the Thatcher administrations of the 1980s. However, the first attacks on what was regarded as over-centralized and insensitive government came from the left through a number of initiatives to make local authorities more accessible and therefore popular with the populations they served. These schemes generally focused on decentralizing services, typically through the establishment of neighbourhood offices. Despite this, undoubtedly the most profound and sustained challenge to the traditional bureaucratic form of working has come from the Conservative governments which have been in power since 1979.

The analysis that underpinned the subsequent legislative changes of these governments was that by exposing the provision of local government services to market forces, quality would be improved while simultaneously reducing waste and inefficiency (see, for example, Osborne and Gaebler 1992). Moreover, a greater choice of service providers would make local authorities more accountable to their local electorates (Flynn 1990). Overall the introduction of competition into service provision was probably motivated as much by a desire to reduce public expenditure as arguments that linked market mechanisms with enhanced quality and accountability. Regardless of the underlying motivations, the main stimulus for organizational change in local government has been the imposition of compulsory competitive tendering (CCT) for the provision of local services. This allows outside organizations to tender, with the organization offering the lowest bid within these terms of reference having the opportunity to deliver the particular service. At first it was simply manual activities such as refuse collection and road maintenance that were affected, but subsequently the tendering process has been extended into professional tasks such as housing management and computing support.

The introduction of CCT has had significant implications for the organization of local authorities. In particular there has been a widespread tendency towards the decentralization of day-to-day budgetary and administrative control through the establishment of cost centres or business units. These organizational units usually consist of a small subsection of a department and are assumed to be better able to compete in an increasingly market-oriented public sector than traditional large and relatively unwieldy departments. However, decentralization is not simply a matter of devolving responsibility internally within the organization but also externally to outside concerns. Decentralization therefore involves both contracting in as well as contracting out, with the balance of internal and external provision depending on the values and approach of the particular local authority. The concept of enablement has been strongly associated with this process. The exact meaning of this term has become increasingly open to question but there appears to be an underlying sense that local authorities should concentrate on the strategic rather than the operational aspects of service delivery. As a result, it is argued that the role of local authorities is to create a framework through which they enable others with greater experience and knowledge of the needs of potential customers to provide the service.

Decentralization is not the only process taking place within local authorities, for while responsibility for the operational aspects of service delivery is being devolved, strategic control is increasingly being centralized. Inherent within a system of working based on contractual relationships must be a means of enforcement. Consequently, performance indicators have become the fashionable means of assessing quality, while efficiency is judged in terms of reduced costs. The adoption of such mechanisms within the local government sector is to a large extent a reflection of their use by central government as a means of asserting control over service provision and most particularly expenditure at the local level. The increased use of performance indicators and their aggregate form in league tables both within the private as well as public sector has been linked to the facilities offered by computers, particularly their ability to store and manipulate vast quantities of data. However, alongside this cult for quantitative measures is a scepticism about the level of insight that such data provides into the actual quality of the service delivered (Postman 1992; Roszak 1994). This can be seen in the struggle between managerialism and professional independence which is occurring in most public sector organizations at the present time. The ongoing battles between clinicians and managers in the National Health Service in Great Britain has gained considerable publicity but similar conflicts can be seen throughout the public sector in many countries. Even where such tensions do not result in overt forms of protest it is often possible to discern a significant difference between the espoused language of performance indicators and the actual values that underpin the work of professionals (Argyris and Schon 1978).

Post-bureaucratic forms of management, as they have been termed by Hoggett (1991), therefore exhibit all the tensions and struggles for autonomy that were inherent within the earlier styles of administration. Phrases that seem somewhat contradictory are now being applied to the style of management evolving in local government, including decentralized centralization, formalized informality, devolution without power and freedoms within boundaries. The dilemma for all forms of organization is how much discretion can be transferred away from the centre without anarchy breaking out. At the same time, if control rests exclusively with the centre, flexibility and a capacity for innovation are likely to be strangled while considerable talent and energy is wasted on covert as well as overt attempts to challenge the existing distribution of power.

Central government's underlying vision of public sector management in the 1980s and into the 1990s appears to be one of discretion within tight and closely monitored boundaries. Many would argue that the areas over which discretion can be exercised are relatively trivial in nature. Nevertheless, as a growing number of case studies are suggesting, how this style of management is made to work in practice depends upon the values and practices inherent within the particular organizational culture (see, for example, Pettigrew, Ferlies and McKee 1992). Consequently, the boundaries that are set and the scope of flexibility that is given to professionals will vary from one context to another. Similarly, if there is a dysfunction between these boundaries and the organizational culture it is likely that considerable disruption will result. As Chapter 2 suggests in relation to organizations in general, management within the public sector will continue to rely on the power of socialization to regulate behaviour.

There appears to have been a change in emphasis in the management style that characterizes the organizations making up the local government sector over the last 20 years. The image, which to a certain extent still persists, of large unwieldy bureaucracies has been replaced by more flexible administrative structures. The use of the term 'post-bureaucratic management' to describe the current situation indicates that the exact nature of this change is not clear as yet. Moreover, there is increasing recognition that the responses of individual authorities to the changes imposed by central government vary considerably. These circumstances, to a large extent, parallel experience in the private and voluntary sectors. Bureaucratic structures are by no means unique to local government as the cumbersome nature of many multi-national corporations bears witness. The review highlights that, regardless of the overall structure, one of the most enduring aspects of organizational life is the constant tension between the centre and operational units, or to put it another way, managerialism and professionalism. In the end, political accountability is perhaps the only real distinction that remains between the private and public sectors, as the non-profit aspect of the latter has increasingly been eroded. Local authorities, like organizations in general, are collections of people undertaking a variety of tasks and it is within this context that the subsequent analysis is set.

Research strategy

It was noted in Chapter 1 that, despite the considerable claims that have been made about the potential of GIS technologies, very little detailed analysis has been undertaken that explores the extent to which these opportunities are being realized in practice. Given the theoretical framework outlined in Chapters 2 and 3, a research strategy was devised which attempts to provide insight into the processes that affect the everyday experiences of users. The following discussion provides an overview of the research strategy which underpins the findings presented in the subsequent chapters. Detailed methodological issues will be considered at the appropriate point in the analysis.

There has been much debate in the general literature on computing as well as within the GIS field as to the most appropriate research methods to adopt in studying the impacts of information technology (see, for instance, Boland and Hirschheim 1987; Campbell 1990a; Kling 1987; Kling and Scacchi 1982; Mumford, Hirschheim, Fitzgerald and Wood-Harper 1985; Onsrud and Pinto 1991; Pettigrew 1988b; Van de Ven and Rogers 1988). In particular, discussion has concentrated on the merits of case studies and the manner in which such studies should be undertaken. Kling (1987), in a useful overview of the field, distinguishes between research strategies based on what he describes as a discrete-entity approach and those adopting a web model. Discrete-entity approaches are said to concentrate on the views of the technical specialists and those most involved with the technology and formal statements of intent, usually in the form of implementation strategies. In contrast the web model emphasizes the importance of exploring the whole social and political network within which computer-based systems are located. The aim therefore is to get behind the formal appearance of activities within a particular organization to the underlying values and practices that influence how things actually operate in practice. Consequently, the boundaries of the two types of study vary markedly. For instance, instead of just interviewing those most involved with the project, web-type approaches stress the valuable insights to be gained from a much wider group including senior managers, potential users and junior members of staff. Moreover, the web model suggests that the research strategy adopted should incorporate some measure of flexibility as it is unlikely that all the variables that will need to be considered can be specified at the start of the research. As a result, rather than relying on closed questionnaire-based methods, the web model emphasizes the importance of using semi-structured or unstructured interviews, participant observation and longitudinal studies.

The research strategy on which the empirical investigations were based combines a variety of approaches, including comprehensive telephone surveys of the 514 local authorities in Great Britain and 12 case studies. The analysis of the case studies addresses the main questions raised in the preceding chapters, particularly the nature of the relationships between innovations and organizations. However, in order to provide the context for the detailed research and assist with case study selection it was important to establish a broad overview of trends within the local government sector. Moreover, as there was little more than anecdotal evidence available as to the extent of actual GIS adoption, the survey of take-up had considerable value in its own right.

Two complete surveys of GIS adoption in British local government have been undertaken: the first between February and June 1991 provided a basis for the selection of the case studies; and the second took place during the summer of 1993 (Campbell and Masser 1992; Campbell, Masser, Poxon and Sharp 1994). In each case a telephone-based survey approach was adopted. This method assured that a 100 per cent response rate was achieved, thereby overcoming the problem of how to interpret the inevitable non-responses from a postal survey. The approach also ensured that there was no ambiguity between respondents as to what constituted a GIS as the interviewer was able to clarify the capabilities of the technology with the respondent. The aim of each interview was to establish whether or not there was a GIS present within the authority and, if so, to collect basic information on its technical and organizational characteristics and the types of benefits and problems that were being encountered. In addition, the adoption of a telephone-based survey approach resulted in valuable subsidiary information being collected. For each separate system present within an authority one respondent, generally the project manager, was interviewed.

While the information collected from the survey provided useful background, it was the case study element of the research that was designed to explore the processes influencing the implementation and utilization of GIS technologies within organizations. Twelve case studies were selected which represented about 12 per cent of the total number of systems present within British local government at the time of the initial survey in 1991. The main phase of fieldwork was conducted during 1991 and 1992, although contact has been maintained with the authorities and in two cases a detailed programme of research has been under way since 1987 (see Campbell 1990a). The longitudinal element to the research has been used to enhance understanding of the underlying processes, while the descriptions of the characteristics of the various systems are based on the circumstances found during the main phase of fieldwork so as to ensure comparability.

The findings of the 1991 survey were used as the basis for the selection of the case studies. Four key criteria were chosen with the objective of highlighting cases that would provide evidence as to the appropriateness of the underlying assumptions concerning the various interpretations of implementation discussed in Chapter 3. As a result a range of hardware and software platforms were chosen to enable the technological aspects of implementation to be assessed. Similarly, a range of organizational approaches to implementation were selected, including departmental and so-called 'corporate' systems to provide the basis for the analysis of a variety of managerial strategies. In addition it was also felt to be important to ensure that a range of local authority types was included and equally that the systems had been present within the particular organization for at least two years prior to the start of the fieldwork in 1991. This criterion was important as it enabled the research to concentrate on the consequences of the implementation process rather than the initial technical problems associated with the introduction of most computer-based systems. All the selected case studies gave their full cooperation to the conduct of the research. The willingness to participate owed a great deal to the initial contact which had been made during the telephone survey. The details of the selected case studies will be discussed following the review of the survey findings in the next chapter. However, the breadth of the environments chosen compares favourably with the largely anecdotal evidence that currently exists concerning GIS implementation (see, for example, James and Pope 1993; Moore 1994; van Buren 1991; Winter 1991).

The approach used in the case studies closely followed the web model described above (Kling 1987; Kling and Scacchi 1982). It was also possible to draw on the previous experience of the research team in undertaking studies of this type (see Campbell 1990a). The length of time taken to complete each of the case studies varied from one day to two weeks depending on the size and complexity of the organizations and the GIS being developed. In most cases an exploratory visit was followed by a period of more detailed fieldwork. A key feature of the approach adopted was that interviews were undertaken with those both directly and indirectly involved in technology in the various organizations. As a result, semi-structured interviews were conducted not only with project managers but also senior managers, data processing staff and mapping and computing specialists. The information gained from the interviews was complemented by general observation of the practices of the organization as the time between interviews was spent in the case study authorities. The objective of this approach was to ensure that a full understanding could be gained of the organizational context into which the process of GIS implementation was embedded. It was not regarded as adequate simply to rely on the formal statements of intent but rather to attempt to uncover how things actually worked within the environments studied. Consequently, GIS were not treated as discrete entities but rather to be embedded within the practices, values, power relationships and conflicting priorities of the organizational contexts examined.

The research strategy therefore combines basic survey evidence with in-depth case study analysis of 12 instances of GIS implementation. The crucial element in shaping the overall approach to the empirical investigations was the development of the theoretical framework discussed in the preceding chapters. As a result this research represents the first systematic attempt to explore the processes influencing the implementation and utilization of GIS technologies in practice. It must be acknowledged that there are limits to the generalizability of any form of case-study-based research (Yin 1982 and 1994). However, while all organizational contexts are inherently unique, the understanding gained from the experiences of the 12 local authority environments provides valuable insights into the relationships between organizations and technological innovations (Bryman 1988).

Overview of the research findings

The following three chapters analyse the findings of the research. Chapter 5 outlines the findings of the two surveys. It seeks to explore the extent to which GIS technologies have been adopted by local authorities in Great Britain, the technological and organizational characteristics of the systems and the expectations and problems that have been encountered. These findings provide the basis for the case study selection and the first indications as to the ease of implementation and type of managerial structure being favoured for the introduction of GIS. This analysis results in a threefold typology of the styles of implementation being adopted in the various contexts. Chapters 6 and 7 examine the implementation process in more detail, based on the case study findings. The first of these concentrates on the extent to which the basic form of the innovation is influenced by the organizational context. The extent to which reinvention is taking place in each context is examined and actual levels of GIS utilization being achieved are discussed. Chapter 7 builds on these findings to explore whether effective implementation and utilization are dependent on the technological characteristics of the GIS, rational managerial strategies or a more complex set of social and political processes. Throughout the discussion the aim is to consider the interaction between an innovation, in this case GIS, and the organizational contexts found in local government so as to provide insights into how users in everyday settings are responding to GIS and what processes account for their reactions.

CHAPTER 5

The diffusion of GIS in British local government

Introduction

This chapter describes and evaluates the findings of the two comprehensive telephone surveys of British local authorities that were carried out in spring 1991 and summer 1993. The greater part of the discussion focuses on the findings of the more recent survey. These are analysed with respect to levels of adoption, system development, choice of technology and perceived benefits and problems associated with GIS. Then these findings are compared with those from the earlier survey in order to highlight the volatile nature of GIS diffusion and to illustrate the dynamics of current developments in local government. The last section of the chapter evaluates the findings of the two surveys in relation to the theoretical concepts developed in earlier chapters. In the light of this evaluation a threefold typology of styles of implementation is proposed for the selection of the case studies discussed in subsequent chapters.

GIS adoption

Table 5.1 gives an overall picture of the state of local authority plans for GIS in summer 1993. From this it can be seen that 149 out of the 514 authorities, or 29 per cent of the total in Great Britain, had GIS facilities and that a further 50 authorities had firm plans to acquire them within a year. Many of these have already set up working parties and undertaken feasibility studies to evaluate GIS. As a result, it is likely that about 2 authorities in 5 in Great Britain will have GIS facilities at the end of 1994.

Some 139 authorities were considering the acquisition of GIS at the time of the survey and only 176 authorities had no plans to introduce GIS. The main

GIS and organizations

Table 5.1 Plans for GIS in local authorities in Great Britain

	1993	
Plans for GIS	Number	%
Already have GIS facilities	149	29.0
Plans to acquire GIS within one year	50	9.7
Considering the acquisition of GIS facilities	139	27.0
No plans to introduce GIS	176	34.2
TOTAL	**514**	**99.9**

Table 5.2 Authorities with GIS by type of local authority

	With GIS	
Type of authority	Number	%
Shire districts	61	18.3
Metropolitan districts	34	49.3
Shire counties	43	91.5
Scottish districts	5	9.4
Scottish regions	6	66.7
Scottish Islands	0	0.0
TOTAL	**149**	**29.0**

reasons given by the latter were the uncertainties associated with the government's proposals to reorganize local government and lack of finance. However, a number of authorities reported that they had set up working parties to consider GIS and decided not to proceed as a result of their negative recommendations.

The findings demonstrate the high level of awareness of GIS in local government circles in Great Britain even among authorities that had no plans to invest in such systems. However, there were marked differences between different types of authority and also between different regions with respect to GIS adoption. Table 5.2 shows the number of authorities that have already acquired a GIS by local authority type. From this it can be seen that the highest level of GIS adoption is at the shire county and Scottish region levels where 43 out of the 47 counties, or 91 per cent of the total, and 6 out of the 9 regions already had GIS facilities.

In contrast only 61 out of the 333 shire districts (18.3 per cent), 5 out of the 53 Scottish districts (9.4 per cent) and none of the three Scottish Island

The diffusion of GIS in British local government

Table 5.3 Percentage of local authorities with GIS facilities by type and region

Type of authority	% South with GIS	% North with GIS	% all authorities
Shire districts/Scottish districts	21.0	10.9	17.0
Metropolitan districts	52.5	44.8	49.3
Shire counties/Scottish regions	90.3	84.0	87.5
All authorities	32.2	24.3	29.0

authorities had acquired GIS by summer 1993. As might be expected, the level of adoption in the metropolitan districts fell mid-way between those of the shire counties and Scottish regions on the one hand and those of the shire districts and Scottish districts on the other. Table 5.2 shows that 34 out of the 69 metropolitan districts (49.3 per cent) had GIS at the time of the survey. It is important to note, however, that despite the relatively low level of adoption in shire districts this type of authority had the largest absolute number of authorities with GIS.

Table 5.2 also shows that levels of adoption in the Scottish regions and Scottish districts are generally lower than those in England and Wales. The extent of these regional differences is further explored in Table 5.3 with reference to the North/South divide. This indicates that, in overall terms, 32 per cent of authorities in the South, which includes London, the South East region together with East Anglia, the East and West Midlands and the South West, had GIS facilities as against only 24 per cent of authorities in the North, which includes Wales, Yorkshire and Humberside, the Northern and North Western regions as well as Scotland. As might be expected, these variations were least pronounced with respect to the shire counties and Scottish regions where GIS adoption is approaching 100 per cent, and most pronounced with respect to the shire districts and Scottish districts where GIS adoption is still under 20 per cent. In the former case the ratio of South to North in percentage terms is only 90:84 whereas it is 21:11 in the latter. Generally, differences between the metropolitan districts are closer in this case to those of the shire counties and the Scottish regions with a ratio of 53:45.

The picture of GIS adoption in summer 1993 that emerges from these survey findings is very clear. Overall levels of adoption were highest in the shire counties and the Scottish regions where nearly all authorities had GIS facilities, and lowest in the shire districts and Scottish districts where less than 20 per cent of authorities had GIS. About half the metropolitan districts had GIS at the time of the survey. Levels of adoption were also high in the Southern and Eastern regions of Britain and lower in the Northern and Western regions. The extent of these regional variations was particularly marked with respect to the shire districts and Scottish districts. In this case the probability that a shire district in the South had GIS was nearly twice as high as that for a shire district or Scottish district in the Northern part of Great Britain.

GIS and organizations

System development

The previous section dealt with the number of authorities with GIS facilities. To analyse GIS usage it is necessary to consider the number of systems purchased by these authorities and their configuration. A system is regarded as a distinct piece or combination of software that one or more departments within a local authority are implementing. For instance, a situation where several departments are developing separate applications based on the same software is considered as one system.

Table 5.4 shows that a total of 195 separate systems had been purchased in 149 authorities which had acquired GIS facilities by summer 1993. Shire counties and Scottish regions were most likely to have more than one system within an authority with 75 systems in 43 authorities and 11 systems in 6 authorities respectively. In contrast only one of the 61 shire districts and 2 of the 7 Scottish districts had more than one system.

Table 5.5 shows that more than 70 per cent of all authorities had purchased systems since the beginning of 1990. The peak year for local authorities to move into GIS for the first time was 1990 itself when 35 authorities acquired facilities. Since that time the number of authorities adopting GIS each year has declined slightly. Table 5.5 also shows that over three-quarters of all systems had been purchased since the beginning of 1990 and that the peak year for system acquisition was 1992. It should be noted, however, that the figures for 1993 are not complete because of the timing of the survey.

The differences noted above regarding the number of systems in various types of authority reflect both the resources at the disposal of the authority and the organizational arrangements within that authority for carrying out their statutory duties. At the same time they give some indication of the diversity of approach that is being adopted to GIS implementation in British local government. From Table 5.6 it can be seen that 105 out of the 195 systems were at the disposal of a single department while the other 90 systems were shared by more than one

Table 5.4 Level of GIS adoption by authorities and systems

Type of authority	No. of authorities possessing GIS	Number of GIS
Shire districts	61	62
Metropolitan districts	34	40
Shire counties	43	75
Scottish districts	5	7
Scottish regions	6	11
TOTAL	**149**	**195**

The diffusion of GIS in British local government

Table 5.5 Length of experience with GIS technologies

	Authorities		Systems	
Year	Number	%	Number	%
Before 1986	5	3.4	7	3.6
1986	1	0.7	2	1.0
1987	3	2.0	5	2.6
1988	15	10.1	16	8.2
1989	12	8.1	15	7.7
1990	35	23.5	39	20.0
1991	26	17.4	36	18.5
1992	30	20.1	46	23.6
1993*	22	14.8	29	14.9
TOTAL	149	100.1	195	100.1

Note: Up to September only

Table 5.6 Approach to GIS implementation

	Single department		More than one department		
Type of authority	Number	%	Number	%	Total
Shire and Scottish districts	27	39.1	42	60.9	69
Metropolitan districts	19	47.5	21	52.5	40
Counties and regions	59	68.6	27	31.4	86
TOTAL	**105**	**53.8**	**90**	**46.2**	**195**

department. Departmental systems were particularly common in shire counties and Scottish regions where they accounted for 59 out of the 86 systems in use. In contrast, systems involving more than one department accounted for 3 out of every 5 facilities in the shire districts and Scottish districts. In the metropolitan districts the distribution of single and multi-departmental systems was relatively evenly balanced with 19 single as against 21 multi-departmental systems.

In no way can multi-departmental systems be equated automatically with corporate systems in the sense discussed in Chapter 3. Table 5.7 shows that the majority of multi-departmental facilities involved only two or three departments and less than one GIS facility in five involved five or more departments.

GIS and organizations

Table 5.7 Number of departments involved in multi-departmental GIS facilities

Number of departments	Shire/Scottish districts %	Metropolitan districts %	Counties & regions %	Total %
2	19.0	23.8	40.7	26.7
3	35.7	19.0	29.6	30.0
4	9.5	14.3	14.8	12.2
5	19.0	14.3	0.0	12.2
More than 5	9.5	28.6	14.8	15.6
All departments	7.1	0.0	0.0	3.3
	(n = 42)	(n = 21)	(n = 27)	(n = 90)

Table 5.8 Departments involved in GIS facilities

Departments	Shire/Scottish districts	Metropolitan districts	Counties & regions	All authorities
Planning/development	49	25	32	106
Highways/engineers/surveyors	16	15	28	59
Estates	19	13	19	51
IT/computer services	14	5	14	33
Combined technical services	12	7	12	31
Legal and related services	19	4	0	23
Parks/recreation	13	5	2	20
Chief executive	6	9	5	20
Others	24	27	10	61
TOTAL	**172**	**110**	**122**	**404**

Facilities involving two or three departments were particularly popular in the shire counties and the Scottish regions where they accounted for over 70 per cent of all multi-departmental facilities. However, even in the shire districts and Scottish districts, facilities involving two or three departments accounted for over half the total in this category. Conversely, only 7 per cent of facilities in the shire and Scottish districts and no facilities at all in the metropolitan districts or the shire counties and Scottish regions involved all departments in the authority.

The diffusion of GIS in British local government

Table 5.9 Lead departments in multi-departmental facilities

Departments	Shire/Scottish districts	Metropolitan districts	Counties & regions	All authorities
Planning/development	17	11	4	32
IT/computer services	4	4	5	13
Combined technical services	4	1	5	10
Highways/engineers	0	0	5	5
Chief executive	3	0	1	4
Legal and related services	3	0	0	3
Others	4	2	2	8
No lead department	7	3	5	15
TOTAL	**42**	**21**	**27**	**90**

Table 5.8 shows that 404 separate departments were involved in the 195 GIS facilities in summer 1993. A quarter of these departments were planning or development departments. Another quarter was accounted for by highways and estates departments. IT and technical services were also well represented, as were legal services, parks and recreation and the chief executive's department. The 'others' category, which accounted for 61 out of the 404 departments, included big-spending departments such as education (10 cases), housing (9) and social services (5), as well as environmental health (14) and buildings and works (4).

In more than a third of multi-departmental facilities, as Table 5.9 shows, planning was the lead department. Only IT/computer services and combined technical services were lead departments in 10 or more facilities and in 15 cases there was no single lead department. Planning was particularly important as the lead department in shire district and Scottish district applications and at the metropolitan district level. At the county level, planning together with highways and the two technical services departments were all lead departments in 4 or 5 instances.

Table 5.10 shows that planning departments also accounted for a third of all single department GIS facilities and over half at the shire district/Scottish district level. Other major users of single department facilities were highways/ engineering and emergency services, especially in the shire counties and the Scottish regions, and combined technical services and estates.

The overall picture of GIS system development that emerges from the survey findings is one of considerable diversity reflecting the specific demands of particular types of authority and different approaches to organization and

GIS and organizations

Table 5.10 Single department GIS

Departments	Shire/Scottish districts	Metropolitan districts	Counties & regions	All authorities
Planning/development	15	8	12	35
Highways/engineers	2	2	12	16
Emergency services	0	0	15	15
Combined technical services	4	2	7	13
Estates	2	2	7	11
Others	4	5	6	15
TOTAL	**27**	**19**	**59**	**105**

management. The findings generally suggest that GIS in British local government is decentralized and largely a bottom-up activity and that centralized systems are generally in the minority. Single departmental systems predominate in shire county and Scottish region GIS facilities while multi-departmental systems were the most common feature of shire district and Scottish district facilities. The tendency towards fragmentation of GIS in the counties which was evident in the number of systems in use is further reinforced by an emphasis on two or three departmental systems even in cases where multi-departmental systems are being implemented.

In all types of authority, planning and development departments were the most frequently cited department. Planning was also the lead department in over a third of the multi-departmental facilities. The predominance of planning reflects the importance attached by these departments to geographic information as well as to their traditional responsibilities for meeting the cartographic needs of local authorities. The pre-eminent position of planning was challenged only by highways/engineering departments in the shire counties and Scottish regions with respect to single department facilities. Given this emphasis on operational departments, it is not surprising to find that IT/computing services together with other central departments such as legal and related services and the chief executive's departments were lead departments only in a relatively small number of multi-departmental applications.

Choice of technology

Table 5.11 provides a breakdown of the software packages acquired by local authorities. It should be noted that systems with GIS capabilities that are largely being used to perform activities such as computer-aided design (CAD) have been omitted from this analysis. It should also be noted that where packages had been

Table 5.11 Software adopted for GIS work

Software	Shire/Scottish districts %	Metropolitan districts %	Counties & regions %	All authorities %
Arc/Info	10.1	27.5	29.1	22.0
Wings	5.8	–	11.6	7.2
MapInfo	8.7	7.5	4.6	6.7
Axis	13.1	10.0	–	6.7
Alper GIS	10.1	10.0	1.2	6.2
G-GP	5.8	15.0	1.2	5.6
X Assist	–	–	11.6	5.1
GDS	8.7	5.0	1.2	4.6
Others	37.7	25.0	39.5	35.9
TOTAL	**100.0**	**100.0**	**100.0**	**100.0**
	(n = 69)	(n = 40)	(n = 86)	(n = 195)

purchased to provide specialist GIS facilities to supplement an existing system, it is the main system that is recorded in the table.

A striking feature of the survey findings is the pre-eminent position of Arc/Info (ESRI) as the leading GIS software package in British local government with a market share of about 22 per cent. Arc/Info was used mainly by the shire counties and Scottish regions together with the metropolitan districts. The market leader for the shire districts and Scottish districts was Axis (a system developed by a British company in Northampton) with 13.1 per cent of the total number of systems. Other software packages used extensively at the shire district and Scottish district level were Alper GIS (another system developed by a British company based in Cambridge which is now part of Sysdeco) and Arc/Info. Arc/Info, Axis and Alper GIS were also used extensively by the metropolitan districts, as was G-GP (another British system). The latter is largely due to the development of the software for the Research and Intelligence Section of the old Greater London Council and its subsequent popularity with the London boroughs. Two packages that were widely used at the shire county and Scottish regional level but not elsewhere were Wings (another British system) and X Assist (a specialist system developed for environmental monitoring).

Table 5.11 also highlights the variety of software that was in use at the time of the survey. The 'others' category accounted for over a third of all applications. This included more than 30 additional software packages as well as some in-house home-grown products. The greatest range of systems was found at both the shire county and the Scottish regional level and in the shire districts and Scottish districts.

GIS and organizations

Table 5.12 Hardware adopted for GIS work

Hardware	Shire/Scottish districts %	Metropolitan districts %	Counties & regions %	All authorities %
Workstation	55.1	60.0	53.5	55.4
Micro	34.8	22.5	34.9	32.3
Mainframe	10.1	7.5	5.8	7.7
Mini	–	10.0	1.2	2.6
Other	–	–	4.6	2.0
TOTAL	**100.0**	**100.0**	**100.0**	**100.0**

In the survey, respondents were asked questions about the hardware used to support GIS. In cases where systems had several means of access, for instance mainframe terminals and microcomputers, it was the highest level of capability, i.e. the mainframe component, that was used in the analysis.

Overall, as Table 5.12 shows, mainframe installations accounted for only 1 facility in 12 in summer 1993. The dominant hardware platform for GIS in all types of authority was the workstation, followed by the PC. These two hardware types accounted for 55.4 and 32.3 per cent of all applications respectively.

Benefits and problems associated with GIS

Respondents were asked to rank in order of importance three sets of benefits and problems associated with GIS. They were then asked to describe in more detail the group of benefits and problems they had ranked first with reference to a more detailed list of topics. Provision was also made for issues not identified in the questionnaire to be raised. Although the findings from this part of the survey are particularly dependent upon the perceptions of the individuals interviewed, they nevertheless enable some general issues to be identified.

Table 5.13 shows that the most important benefit associated with GIS was better information processing. More than 60 per cent of all respondents ranked this first. Their main reasons for giving this first place were improved data integration, increased speed of data provision, better access to information and an increased range of analytical and display facilities.

The next most important factor in the ranking was better quality decisions which was placed first by 20.8 per cent of respondents. Better quality decisions were felt to be particularly important by shire counties and Scottish regions. The most important reasons given for this ranking were related to operational and managerial decision-making rather than policy-making matters. The third factor, general savings, was placed first by only 11.4 per cent of respondents. In this

Table 5.13 Most important benefits associated with GIS

Benefits	Percentage of all authorities in 1993
Improved information processing facilities	61.4
Better quality decisions	20.8
General savings	11.4
Others	6.4
	($n = 98$)

Table 5.14 Most important problems associated with GIS

Problems	Percentage of all authorities in 1993
Technical	29.0
Organizational	27.2
Data related	26.3
Others	7.6
No problems	9.8
	($n = 195$)

case the most important savings were associated with reductions in time rather than in cash or staff.

Table 5.14 shows that respondents were more or less evenly divided between organizational, technical and data-related issues when it came to ranking problems. A wide variety of technical issues were raised by respondents who ranked technical matters first in their list of problems. The most commonly cited technical problem was hardware reliability, followed by lack of systems compatibility. Other problems cited by respondents included software capability, lack of user-friendliness and difficulties with vendors.

The most important organizational problems perceived by respondents were the poor quality of managerial structures for implementing GIS, followed by shortage of skilled staff and the lack of encouragement from senior staff. In several cases staff resistance also presented a significant problem.

By far the most commonly cited data-related problem was the cost of data capture. Respondents were not only concerned about the cost associated with data capture but also the amount of time required for this task. Other problems cited by some respondents included lack of compatibility between data sets, poor quality and difficulties with Ordnance Survey data. Problems of lack of

GIS and organizations

compatibility were predominantly cited by respondents from shire counties and Scottish regions.

Main changes since 1991

The 1993 survey was carried out in the same format as that used in a survey undertaken two and a half years earlier during spring 1991. Because of this it is possible to compare directly the findings of the two surveys to highlight the dynamics of GIS diffusion within British local government. Some of the main similarities and differences between the surveys are summarized in Tables 5.15 and 5.16 with respect to the headings used above.

From Table 5.15 it can be seen that the number of authorities with GIS had increased by 75 per cent over the two and a half years from 85 to 149. As might be expected, the rate of increase was lowest in the shire counties and Scottish regions where adoption levels were approaching 100 per cent, and highest in the shire districts and Scottish districts where adoption levels were still below 20 per cent. In the latter the number of authorities with GIS had more than doubled since the time of the 1991 survey.

These figures highlight the volatile nature of GIS diffusion in British local government over the last few years. If the expectations of the 50 additional

Table 5.15 Summary of main changes between 1991 and 1993 with respect to GIS adoption and system development

	1991	1993
Adoption		
No. of authorities with GIS	85	149
Shire counties/Scottish regions	35	49
Metropolitan districts	22	34
Shire districts/Scottish districts	28	66
South/North ratio (% adoption)	20:11	32:24
System development		
No. of systems in local authorities	98	195
Average length of experience (years)	2.51	2.94
Single/multiple departmental ratio	49:51	54:46
Average no. of departments in multi-departmental applications	4.18	3.81
% planning as lead department in multi-department applications	27.0	35.6
% planning of single departmental applications	22.9	33.3

authorities that indicated that they had firm plans to acquire GIS within one year have since been realized, the overall level of GIS adoption would have risen by a further 35 per cent over 1993 levels by the end of 1994. In practice, even this could be an underestimate given the boost to GIS adoption given by the Service Level Agreement reached between the local authorities and between Britain's national mapping agency, the Ordnance Survey, and the local authority associations in March 1993 regarding the acquisition of topographic information in digital or geographic format. This marks an important departure from previous agreements whereby individual local authorities or departments have negotiated their own terms with the Ordnance Survey for the purchase of such data. It is also an indication of the increasing availability of large-scale digital map data in Great Britain as the Ordnance Survey completes its comprehensive digitization programme for the country as a whole.

With this in mind it is worth noting that 460 local agreements had been signed by local authorities by the end of the first year of operation in April 1994. Some 289 of these authorities, or 62.8 per cent of the total, opted for large-scale digital map data. This number is nearly double the number of local authorities that had GIS facilities in September 1993 (Campbell *et al.* 1994, p. 37).

Table 5.15 also shows that, as overall levels of adoption rise, regional variations tend to be reduced. Whereas the probability of an authority in the southern part of Britain adopting GIS was nearly two to one in 1991, by 1993 it was only four to three. However, these overall figures conceal important differences between authority types. Between 1991 and 1993 the South:North ratio actually increased slightly with respect to the smaller shire districts and Scottish districts whereas it was already approaching one to one for the shire counties and Scottish regions.

During the two and a half years between the two surveys the number of systems in local authorities had effectively doubled from 98 to 195. However, because of the rapid increase in acquisitions, the average length of experience with these systems had risen only from 2.51 to 2.94 years during the period. Consequently, for the vast majority of British local authorities, GIS is still a new technology and they are still in the process of building up operational experience.

The growing fragmentation of GIS in British local government which is evident in the number of authorities that have acquired more than one system, especially at the shire county and Scottish region level, is also apparent in the configuration of systems. Table 5.15 shows that the ratio of multi-departmental to single departmental facilities was reversed between 1991 and 1993 from a slight majority in favour of multi-department facilities to a clear 54:46 majority in favour of single department facilities in 1993. It is also worth noting that the average number of departments involved in multi-departmental facilities went down from 4.18 to 3.81 over this period.

Given the enormous diversity of local government operations it is surprising to find that planning and development departments have actually consolidated their position as GIS leaders within local government since 1991. In 1993

GIS and organizations

Table 5.16 Summary of main changes between 1991 and 1993 with respect to GIS technology and perceived benefits and problems

	1991	1993
Technology		
Software packages (% total)		
Arc/Info	22.4	22.0
Wings	4.1	7.2
Axis	7.1	6.7
MapInfo	3.6	6.7
Alper GIS	12.2	6.2
G-GP	7.1	5.6
GFIS	10.2	3.0
Hardware configurations (% total)		
Workstation	40.8	55.4
Mainframe	25.5	7.7
Benefits (% total)		
Improved information processing	60.5	61.4
Better quality decisions	31.5	20.8
Problems (% total)		
Technical	28.4	29.0
Organizational	28.4	27.2
Data related	34.0	26.3

planning and development departments acted as the lead department for 35.6 per cent of multi-departmental facilities as against only 27 per cent in 1991, and planning and development departments were also responsible for a third of all single department applications in 1993 as against only 22.9 per cent in 1991.

Table 5.16 summarizes the main changes that took place in the two and a half years between the 1991 and 1993 surveys with respect to technology and perceived benefits and problems. From this it can be seen that there was a dramatic fall in the proportion of mainframe installations during this period and an increased dominance of workstation-based facilities. In contrast, Arc/Info has retained its market share over this period despite the growing number of software packages on the market. In practice Arc/Info has consolidated its position at a time when the market shares of the other leading packages have declined. This is particularly evident in the case of GFIS whose market share slumped from 10.2 per cent in 1991 to only 3 per cent in 1993.

It is also worth noting that several packages increased their market shares between 1991 and 1993. The main gainers were Wings and MapInfo whose shares rose from 4.1 per cent to 7.2 per cent and 3.6 per cent to 6.7 per cent respectively. It should be borne in mind, that, in addition to first-time purchases, a number of authorities reported that they had switched software during this period. The most commonly cited switch was from GFIS to Arc/Info.

By comparison with these changes there were no dramatic shifts in the benefits and problems associated by respondents with GIS. Improved information processing was perceived as the main benefit by three out of five respondents in both 1991 and 1993. However, the proportion of respondents who chose better quality decisions as the main benefit declined from over 30 per cent to 20 per cent during this period. This may reflect the increased number of district level as against county level facilities.

The only discernible shift in perceived problems was the fall in the number of respondents who saw data-related issues as the main problems associated with GIS. This probably reflects the extent to which the lead time for data capture has been reduced as authorities build up their operational experience as well as the impact of the Service Level Agreement negotiated by the local authority associations with the Ordnance Survey.

Evaluation

The findings of the two surveys must also be evaluated with reference to some of the characteristics of the theoretical perspectives outlined in Chapter 3: technological determinism, managerial rationalism and social interactionism. Of particular importance in this respect is the propensity for adoption and the processes of GIS diffusion that can be seen from the survey findings. GIS adoption is seen to be inevitable from both a technological determinist and a managerial rationalist perception once the inherent technical and managerial benefits to be derived from GIS are recognized. In contrast the outcome is much less clear cut from a social interactionist perspective due to the uncertainties that surround the decision to adopt GIS.

With these considerations in mind it is also useful to explore the extent to which the survey findings reflect some of the key features of the diffusion process described in Chapter 1. There is a great deal of evidence from other research on the diffusion of technological innovation which suggests that the process over time can typically be represented by an S-shaped or logistic curve (see, for example, Rogers 1983 and 1993). In essence this implies that the initial rate of adoption of technological innovations is likely to be relatively slow until a critical mass of users is achieved, after which the rate of adoption increases very rapidly and the direction of the curve moves away from the horizontal axis towards the vertical axis. As time progresses and levels of adoption near saturation the curve tapers off again towards the horizontal axis.

From Table 5.5 it can be seen that the rate of adoption of GIS by British local authorities up to 1991 is broadly similar to the first section of the S-shaped curve,

GIS and organizations

while the subsequent increase in the number of authorities adopting GIS has all the characteristics of the second section of the curve. However, the point at which the curve levels off again to the third section has not yet been reached and therefore it is not possible to predict the final outcome of the diffusion process. From a technological determinist or managerial rationalist perspective it should be 100 per cent, given the inevitability of adoption, whereas from a social interactionist perspective it should fall short of 100 per cent. Consequently it is worth noting in this respect the reference in the section 'GIS adoption' of this chapter (p. 66) to the fact that a number of authorities reported that they had set up working parties to consider GIS and decided not to proceed as a result of the negative recommendations. Findings such as these cast some doubt as to the inevitability of GIS adoption in all British local authorities.

It is also useful at this stage to introduce another component of diffusion research which gives insights into the spatial dimension of the diffusion of technological innovations (see, for example, Hägerstrand 1952 and 1967). There are two basic spatial models that are of interest in this respect. The first of these is the hierarchical model which postulates that adoption will begin in the larger centres/organizations and subsequently diffuse to smaller centres/organizations. Large centres are seen as being more open to the outside world because of their size and more likely to take on the role of pioneers because of the resources at their disposal. Conversely it is claimed that small centres are less likely to be aware of new developments and also less likely to have the necessary resources to experiment with new tools.

The second spatial diffusion model is the core–periphery model. Like the hierarchical model, this assumes that adoption will begin in core cities or regions because of their size and their links to the outside world, and will then spread outwards to peripheral cities or regions which are less cosmopolitan in character and have fewer resources at their disposal.

Although the survey evidence is essentially cross-sectional in nature, it nevertheless contains many findings that support the assumptions underlying both these models. As far as the hierarchical model is concerned, over 90 per cent of English counties which form the top tier of local government already had GIS facilities at the time of the 1993 survey. In contrast only half the metropolitan districts that form the next tier of local government in terms of size had acquired GIS and less than one in five shire districts and only one in ten Scottish districts that make up the lowest tier of local government had such a facility.

A more surprising finding from the surveys is the extent to which the evidence supports the core–periphery model of diffusion even in a relatively small country such as Britain where there are well-established institutions in existence to disseminate new ideas to all parts of the country at the same time. Nevertheless, as Table 5.3 shows, the probability of local authorities of all types having GIS facilities was higher in the South of Britain than in the North and Scotland. As might be expected, given the findings of the hierarchical model, these differences were least marked at the shire county and Scottish region level and most marked at the shire district and Scottish district levels. Consequently, as Masser's (1993) multi-nomial logit analysis of the findings of the 1991 survey showed, more than half the variance in the data set as a whole can be explained by three variables: type of authority, population size and location in the core or periphery.

The findings of the survey also throw some light on the extent to which technological determinist, managerial rationalist or social interactionist perspectives are reflected in the types of GIS implementation under way and the perceptions of the problems and benefits associated with GIS. With respect to the former, the relatively large proportion of single department facilities, particularly in the counties and regions, which can be seen from Table 5.6, suggest a lack of enthusiasm for information sharing and a strong departmental as against a corporate ethos in many authorities. Furthermore, the extent to which most multi-departmental facilities involve only two or three departments, as shown in Table 5.7, and the relatively small number of facilities involving five or more departments also casts doubt on the extent to which the corporate ideal embodied in the managerial rationalist perspective is being implemented in practice.

With respect to the latter it can be seen from Table 5.14 that GIS implementation has brought with it a variety of organizational, technical and data-related issues as well as the benefits listed in Table 5.13. It is not apparent from these findings whether the benefits clearly outweigh the problems associated with GIS implementation, as might be expected given a technological determinist or managerial rationalist perspective, or whether the outcomes from this comparison were uncertain and varied considerably between authorities, as might be expected from a social interactionist perspective. These issues will be explored in greater depth in the context of the case study research reported in Chapter 6.

A typology of system implementation

Overall, the analysis of the survey findings indicates the diversity of approaches that are being adopted towards GIS implementation in British local government. However, given the distinctive nature of particular systems, a threefold typology of system implementation can be developed from the survey findings (Campbell 1993). These are as follows:

1. *Classically corporate:* The traditional understanding of a 'corporate approach' is reflected in this style of system implementation. Consequently a large number of departments, and possibly the whole authority, participate in the project with the lead taken by the central computer services or planning department. Planning in this case is taking a central coordinating role, probably linked to its existing responsibility for the provision of Ordnance Survey maps or socioeconomic information. The results of this approach are reflected in a workstation- or mainframe-based system using software that is designed to provide limited automated mapping or facilities management capabilities. The choice of software indicates that even in this

context GIS is not primarily expected to enhance the information processing facilities available within the authority. However, in attempting to implement these systems, authorities encounter considerable difficulties. These problems are often technical in nature and not assisted by a shortage of skilled staff.

The findings of the 1993 survey suggest that a classically corporate style of approach has been adopted as the basis for implementing about 15 per cent of the systems found in British local government. This approach is most likely to be favoured by shire districts.

2. *Theoretically/pragmatically corporate:* This style characterizes the approach adopted to implementation which, while involving several departments, does not accord with the classical definition of a corporate approach. This approach is typified by a 'bottom-up' demand for GIS facilities from service delivery departments, or an attempt by the computer service department to achieve some measure of corporate working. On the basis of the survey findings it is difficult to distinguish between approaches that reflect a pragmatic decision among departments to pool resources to facilitate the purchase of equipment and attempts at a coordinated and wide-ranging structure for GIS implementation which have as yet fallen short of this goal. Consequently such approaches are characterized by the involvement of only three or four departments with the lead taken by computing, a technical service department or in certain cases joint responsibility for the project. The systems tend to be workstation or mainframe based but there are no clear trends in terms of software. The main benefits of introducing a GIS are perceived to be enhanced information processing facilities, although some systems in some county councils are expected to improve operational decision-making. Given these objectives, a wide range of problems are being experienced, especially in relation to data and organizational issues.

The findings of the 1993 survey suggest that theoretically/pragmatically corporate approaches are favoured for the implementation of about 35 per cent of systems. Such an approach can be found in all categories of local authority but most especially the metropolitan districts and shire counties.

3. *Fiercely independent:* This approach to implementation is typified by the introduction and development of a GIS by a single department. The department is likely to be involved with technical service type activities such as highways and often has considerable experience in information handling, including the presence of in-house technical expertise. Most of these systems are based on a micro or workstation with a range of software being employed, with a slight tendency towards Arc/Info. It is expected that the GIS will provide improved information processing facilities. However, this goal is inhibited by a number of data-related issues. In addition, organizational problems have not been fully overcome.

The findings of the 1993 survey suggest that a fiercely independent style of implementation appears to typify the approach adopted in about 50 per cent of the systems that are being introduced. Such an approach is most often found in counties and regions.

Conclusion

The findings of the survey show that British local government has already adopted GIS to a very considerable extent. At the same time they raise considerable doubts as to whether the characteristics of the systems currently being implemented by local authorities in practice correspond with the advice contained in much of the GIS literature regarding the importance of information sharing and the need for a corporate approach to GIS implementation. However, it must be borne in mind that the findings of surveys such as those discussed in this chapter can only provide broad indications of the issues involved in the implementation of GIS. In order to understand better the issues involved and the processes that work in these authorities, it is necessary to turn to the findings of the case study research reported in Chapters 6 and 7.

CHAPTER 6

Reinvention and utilization: GIS in practice

Introduction

The findings of the survey component of the research provide a general description of the take-up of GIS technologies over time within one grouping of organizations, namely British local government. The analysis suggests that the implementation of these systems may be more problematic and complex than perspectives based on technological and rational assumptions suggest. As a result these findings raise important questions concerning the nature of the underlying processes influencing implementation within organizations which can only be examined through in-depth investigations. These issues were explored through the case study component of the research and it is these findings that provide the focus for this and the following chapter. Reference will be made throughout to the earlier theoretical discussion, particularly the extent to which the findings from practice provide support for the various assumptions underlying the three perspectives on implementation, namely technological determinism, managerial rationalism and social interactionism.

The presentation of the case study findings has been divided into two parts. This chapter concentrates on the underlying reasons that stimulated adoption of the technology and seeks to describe the technological and organizational characteristics of the GIS being implemented. It also examines the type of applications that were being developed and the extent to which the systems were actually being utilized. The final part of the chapter considers the applicability of the concepts of reinvention and success and failure to the circumstances found in the organizations studied. Are innovations such as GIS independent of the organizational context in which they are located or socially constructed by it? Moreover, of what value are terms such as 'success' and 'failure' in relation to

Reinvention and utilization

the implementation of computer-based systems? Chapter 7 develops this analysis by investigating the relationship between the various organizational contexts studied and the outcomes previously noted. The discussion therefore focuses around the question of what processes influence the utilization of technological innovations? However, before examining the findings the next section provides an overview of the case studies.

The case studies

The criteria used for the selection of the case studies were outlined in Chapter 4. These criteria were subsequently developed in the light of the survey findings. The technical and organizational characteristics of the various systems were summarized in terms of the threefold classification of styles of GIS implementation described in Chapter 5. This typology, alongside a minimum length of experience in handling GIS technology of two years, provided the basis for the selection of the case studies. The use of the typology ensured that a range of types and styles of organization were included. It was also possible to include in the sample contexts that both appeared typical of a particular style of approach and also that deviated from the norm in some important respect. The main characteristics of the case studies are presented in Table 6.1. The greater knowledge about the implementation of the systems acquired prior to conducting the research enabled a distinction to be made between theoretically and pragmatically corporate approaches.

Given the association in much of the literature between GIS technologies, a corporate approach and the benefits of data-sharing, four *classically corporate* systems were included. Two of these closely mirror this style of implementation while the other two provide the additional interest of a *classically corporate* system experiencing organizational problems and a very small authority implementing a large multi-purpose system. *Theoretically corporate* approaches are provided by two metropolitan districts. The large metropolitan-type Scottish authority offers an interesting example of a context in which there were both *pragmatically corporate* and *fiercely independent* systems present within the same organization. Two further *pragmatically corporate* styles of GIS implementation are provided by shire counties. The development of a departmental system by one of the community service-type activities in one of these raises important questions. This system had been purchased more recently than June 1989, but given the history of the authority it was important to examine the issues surrounding the development of this system. In addition to the two departmental systems cited above, a county highways system provides an example of a *fiercely independent* style of GIS implementation which was experiencing organizational difficulties.

As a result of this process, 12 case studies were chosen which represent around 32 per cent of the systems purchased before June 1989, plus one more

GIS and organizations

Table 6.1 Characteristics of the case studies

Style of implementation	Type of local authority	Location (N/S)*	Population (000s) (1991)	Date of GIS purchase (mth/yr)	Majority party (1991)**
Classically corporate	Shire district	S	130	01/89	CON
Classically corporate	Shire district	S	120	06/84	CON
Classically corporate	Shire district	S	80	11/87	LAB
Classically corporate	Shire district	S	30	09/88	IND
Theoretically corporate	Metropolitan district	S	140	03/88	LAB
Theoretically corporate	Metropolitan district	S	130	12/88	CON
Theoretically corporate	Shire county	S	1,500	11/86	CON
Pragmatically corporate	Scottish district	N	650	06/87	LAB
Pragmatically corporate	Shire county	S	1,000	06/86	CON
Fiercely independent	Scottish district	N	650	11/87	LAB
Fiercely independent	Shire county	S	1,500	02/90	CON
Fiercely independent	Shire county	N	950	06/87	CON

* Location: N – North; S – South
**Majority party: CON – Conservative; LAB – Labour; IND – Independent

recent system which was included for the reasons outlined above. It must not be assumed that the 12 case studies chosen are necessarily representative of the experiences of all organizations. However, given the criteria adopted, it was possible to ensure that a range of environments and styles of implementation were included. Moreover, despite the pioneering status of some of the organizations selected it is probable that the responses of users to GIS technologies in these contexts have implications beyond the confines of the environment studied. The experiences of pioneers are also responsible for shaping attitudes towards GIS technology more generally as well as influencing accepted practice.

Why do organizations adopt GIS?

Analysis of the 1991 survey findings indicated that around 50 per cent of GIS take-up in Great Britain can be accounted for on the basis of three characteristics (Masser 1993). These are the size of population served by the authority (greater than 150,000), the type of authority (a county or region) and the authority's location (southern half of Great Britain). Table 6.1 shows that six of the case studies have populations greater than 150,000, four are county authorities and nine are located in the South. Overall, therefore, these authorities appear to be representative of general trends. However, such factors only represent a means of description which helps to summarize what in effect is a mass of independent decisions. This discussion aims to explore the reasons that organizations go to the considerable expense and upheaval involved in introducing a new form of computer technology.

The three perspectives on implementation developed in Chapter 3 provide a range of explanatory frameworks on the reasons that innovations are adopted. The technological determinist and managerial rationalist positions both envisage adoption to be inevitable, given the identification of an operational deficiency. Consequently, a technology such as GIS would not be introduced into an organization without an area of application having already been identified. In contrast the social interactionist approach stresses that the reasons behind the adoption of technological innovations are seldom as clear cut as this. Rather, such actions are likely to reflect an attempt to enhance status and power through the symbolic association of many innovations with progress and rationality. It is also stressed that formal statements providing the official justification for the investment in new technology often differ from the actual motivations that prompted the expenditure. The aim of the case study investigations was to gain insights into the less explicit reasoning behind the purchase of GIS technologies in the organizations studied.

The findings of the case studies suggest the stimulus for GIS adoption to be largely symbolic and political in nature. Only in three cases was the overriding motivation for the introduction of a GIS its ability to alleviate a current organizational problem. In the cases where a positive decision was made to acquire some form of GIS technology, the perceived need was recognized at a departmental level and then purchased and implemented solely by this organizational unit. In two out of these three cases the GIS was purchased as part of a package, partly funded from central resources, to assist with preparations for compulsory competitive tendering for grounds maintenance services. In the third case GIS technology was eventually purchased as part of the process of rationalizing the department's information resources. The reorganization of these facilities had been started prior to the widespread availability of commercial GIS software and therefore the development of such a technology took place by default. This instance, more than any other, provides an example where the task determined the technology to be utilized. However, while in each of these cases an operational need provoked the adoption of a GIS, it was emphasized by respondents that the process of gaining organizational acceptance of this requirement was critical to achieving funding. This was regarded as depending on the tactical skills of those involved as much as the validity of the perceived need. Paradoxically it is the departmental systems that exhibit the closest approximation to the rationally based explanations for adoption, despite this form of implementation being considered by the technological determinist and managerial rationalist perspectives as the least likely to deliver improvements in efficiency or the quality of decisions.

The remaining nine systems demonstrated a less direct relationship between GIS adoption and the main area of application. In some cases the technology was purchased prior to a specific need having been identified. In others, while a formal statement justifying the need for a GIS had been prepared, the actual motivation behind the acquisition of the technology appeared more complex than these documents suggest. Two further groupings of reasons were identified as accounting for the take-up of GIS in the remaining organizations. The first of these concerned a critical aspect of the organizational culture of the authority, while the second stems from the symbolic role of computer-based technologies.

In two authorities there was a strong intellectual tradition in information management, especially spatial data handling. As a result there was a sense of inevitability about the acquisition of a commercial GIS. In many ways both these authorities had been developing skills in geographic information-handling technologies for at least a couple of decades, although their respective strategies were different. Justification for the expense was in both cases based on operational needs, namely improving the capacity of each authority to manipulate Ordnance Survey maps and to manage land and property. However, while these operational needs provided the final grounds for requesting funds, the critical condition that stimulated the take-up of GIS was the existing culture, most particularly a sense of innate innovativeness within the organizations.

In the remaining organizations the acquisition of GIS was strongly linked to the symbolic value of the technology in political and social terms within the organization. This took a variety of forms including, for instance, a new senior member of staff marking a change in approach from their predecessor with the purchase of new computing technology or as a means of a department asserting its independence from the rest of the organization. Whatever the explanation, the introduction of the GIS was used to signal to groups inside as well as outside the authority that a new and progressive approach had been adopted. It can be argued that such actions have a more profound role of attempting to reinforce power and control within an organization. In such circumstances the image of the technology within society, and perhaps more particularly professional associations, is likely to be crucial to the level of adoption achieved.

In five of the authorities a review in the mid-1980s of their computing needs led to the wholesale introduction of information technology. In all these cases much of the data to be held on computer had a spatial component and therefore the respective suppliers who were keen to promote this relatively new product recommended the purchase of GIS software. As a result they acquired GIS technology as a part of a massive investment in computing equipment which never had to be separately justified. This accounts for how most of the small authorities managed to finance the purchase of the software.

The impetus for the introduction of information technology into these authorities largely seemed to rest with the appointment of a new chief executive. In a few cases this reflected a change of approach which was itself initiated by elected members, while in the rest they responded to an initiative brought to them by professional staff. In one case elected members were entirely responsible for requesting the review of computing resources which resulted in the acquisition of a GIS. The strong link between the appointment of a new senior member of staff and expenditure on information technology is not surprising, as the individual concerned attempts to make their mark while elected members demonstrate support for their new appointment through the provision of resources. In the case of a well-established computing environment GIS may itself perform this symbolic function. This occurred in one of the case studies. In this instance the appointment of a new chief executive whose objective was to develop a strong corporate approach coincided with a high-profile marketing campaign by a GIS vendor which targeted the authority.

The adoption of GIS technology may not, however, simply reflect an attempt to signal change. In one of the case studies it was used to symbolize both technical as well as political independence. In this instance the development of the GIS was at the centre of two battles: first, between the two technical service departments and the central computing department, and second, between the two technical departments themselves. While the circumstances are not unique, as the exploration of the process of implementation in the next chapter demonstrates, they were in this instance more overt than is often encountered. This case exemplifies the tension that frequently exists between computing specialists and user departments. The unease reflected a long-established sense of injustice among the technical service departments that their interests had lost out to those of finance-oriented systems. The availability of commercial GIS therefore provided them with a new and technically innovative technology which focused on their land and property-related interests. Interdepartmental rivalries have a significant influence on the development of GIS and often represent the prime motivation for the acquisition of such systems. However, while the motivation may be symbolic, the formal procedures follow the traditional pattern of reasoned justification.

Three groups of reasons have been identified as accounting for the adoption of GIS technology in the case study authorities: perceived need; an innately innovative organizational culture; and the symbolic role of information technology. In four of the cases examined, adoption was facilitated by the financial benefits associated with becoming a 'test site' for a particular vendor's software. The detailed implications of test site status will be explored later in the chapter, although it appears that such activities facilitate the adoption decision rather than determine the outcome.

In addition to exploring the reasons behind GIS adoption the investigations also examined the sources from which organizations gained information about the existence and potential of GIS technologies. The findings indicate three levels of awareness within the case studies. First, in some authorities there was a high level of understanding and knowledge about developments in information processing in general as well as with respect to GIS. This was exemplified by several individuals within the authorities with a considerable tradition and expertise in handling spatial data. The individuals concerned had close links with academic research both through personal contacts and reading specialist journals. Second, in two cases there was evidence of a single individual who, through a chance personal contact, had gained an awareness of the potential benefits of GIS. Finally, in the remaining two-thirds, vendors had been responsible for introducing staff to the potential of GIS technologies. In most cases this was the same vendor from which a system was eventually purchased. Many of the smaller authorities were unaware at the time of purchase that they were adopting a GIS. The role of vendors in stimulating awareness is therefore highly significant. The concentration of vendors in the South East may go some way to account for the relatively slower take-up of GIS in the North of Britain, noted in Chapter 5.

The findings of the research indicate that the simple cause-and-effect model of a problem stimulating the search for a solution which leads in turn to the purchase of an innovation such as GIS has only limited applicability. It is too simplistic to assume that the applications being developed directly reflect the initial motivations that led to the investment in the technology. Moreover, these findings suggest that for an innovation to be adopted it is not sufficient for it to be technically worthy. In many cases the symbolic image of the technology was far more important. In such circumstances the overriding motivation is the short-term objective of purchasing the system rather than the far longer-term goal of utilization. Consequently, to possess a GIS may in some cases be as important as actually using it. Contrary therefore to mechanistic explanations, the *post hoc* search for applications is not due to the stupidity of users but because the initial motivation for the acquisition of the system was not simply stimulated by a perceived need. Overall, these findings tend to support the assumptions underlying the social interactionist perspective. In the end, for most organizations the adoption of GIS technology is a more or less thoroughly reasoned act of faith based on information provided by vendors and motivated by concerns that have little to do with the task that the system will eventually perform.

Characteristics of the GIS technologies being implemented in the case study authorities

The previous section has identified the key reasons that appear to explain the individual decisions made by the case study organizations to purchase a GIS technology. The remainder of this chapter is devoted to describing the technical and organizational characteristics of the GIS being implemented by the case studies. This includes a review of the hardware and software, the people involved, the organizational structures and the applications that were being developed including the types of digital and attribute data that had been input into the various systems. The final part of the chapter examines the extent to which these systems were being utilized and evaluates the findings in terms of the theoretical framework. It should be emphasized that the observations are based on the situation in each of the case studies at the time that the fieldwork was conducted.

Technical characteristics

Table 6.2 indicates the software and hardware configurations utilized by the case studies. In terms of software, 9 different systems were purchased by the 12 case studies, with no particular variations in emphasis according to the size of organization or the style of implementation. For instance, the smallest organization was employing a mainframe GIS, while the departmental systems range from PC-based software through to large-scale systems based on products such as Arc/Info. All the case studies were utilizing the original GIS software they had acquired, with the exception of one of the county councils which had bought a small micro-based system in order to assess the potential of the technology and gain experience before making a more substantial investment. It was, however, likely that two authorities would change their software in the near future, due in

Table 6.2 Technical characteristics of the GIS adopted by the case study authorities

Style of implementation	Software	Hardware
Classically corporate	Axis Amis	Workstation
Classically corporate	McDonnell Douglas GDS Maps	Mainframe
Classically corporate	McDonnell Douglas GDS Maps	Mini
Classically corporate	IBM GFIS	Mainframe
Theoretically corporate	Hoskyns G-GP	Micro
Theoretically corporate	Wang GDMS	Mini
Theoretically corporate	IBM GFIS	Mainframe
Pragmatically corporate	Coordinate	Mainframe
Pragmatically corporate	IBM GFIS	Mainframe
Fiercely independent	Alper GIS	Micro
Fiercely independent	PC Arc/Info	Micro
Fiercely independent	Arc/Info	Workstation

one case to the vendor withdrawing support from the product and, in the second, dissatisfaction with both the software and the support provided by the vendor exacerbated by changes in the local government context. There were also signs from one of the most mature GIS environments that the organization might in the future purchase several GIS products. In this instance the planning department had purchased a separate micro-based GIS to add to the existing mainframe system, specifically in order to handle the 1991 census. It was envisaged that, once customized, the reasonably simple-to-use MapInfo software would provide professional planners with readily accessible mapping facilities with which to interrogate the census. This software had been added to the existing system and therefore was not regarded as separate GIS.

There are some differences in the detailed nature of the equipment being utilized by the case studies as against more recent patterns of take-up. As they were all relatively early adopters their choice of equipment was more restricted than those that followed. As a result, IBM's GFIS software is over-represented, as are mainframe computers as the main hardware platform. Changes in the processing power of computers over the last few years have also blurred many of the differences between different forms of hardware. Consequently, most of the mainframe-based systems have subsequently changed to workstation or even microcomputer-based technologies.

A factor that contributed to the choice of software in four of the case studies was an agreement with a vendor to become a 'test site' for their particular GIS product. The stimulus in all cases was the favourable financial terms on which GIS equipment was made available. For instance, in some cases workstations were leased free of charge to authorities and have subsequently become a permanent part of the equipment of that organization. The vendors, on the other hand, gained access to a real-life environment in which they could develop and refine commercial applications and also utilize these contexts as a shop window through which they could promote their products. In theory, therefore, test site status should be mutually beneficial. However, evaluations of this experience varied considerably, with a small authority suggesting that it had enabled them to achieve results that would not otherwise have been possible while a much larger authority was highly critical of the amount of staff time that they had found it necessary to devote to the project without any similar commitment on the part of the vendor. The software in the latter case has since been abandoned and the authority is extremely wary of test site agreements. The two remaining case studies perceived the consequences of test site status to have been mixed. On the positive side it enabled staff to learn about GIS and acquire the necessary skills to handle such systems at little expense. Furthermore, it was generally felt that the systems adopted were no worse than the others on the market which would have cost a great deal more and would still have required considerable in-house support. Overall financial inducements, whether as part of a 'test site' arrangement or not, certainly had an impact on the assessment of particular GIS products.

Table 6.3 Number of terminals able to access GIS technology

Number of terminals able to access GIS facilities	Number of case studies
1–5	7
6–10	1
11–15	0
16+	4

An important argument that has been used to sell GIS to organizations like local authorities has been the ability of such technologies to increase the availability of key data sets. For this to be, at the very least, physically possible a large number of terminals need to be able to access the facilities throughout the organization or, in the case of a large department, in each separate section. Table 6.3 demonstrates that, on the whole, the case studies appear to fall into one of two groupings. The majority had fewer than 5 specialist terminals while in four cases GIS capabilities were available at more than 16 points within the organization. In 3 of these 4 the seemingly widespread accessibility was due to the GIS software being available through the authority's or a department's computer network. However, one authority had around 20 specialist terminals devoted to GIS-related activities. It is likely that some case studies were starting with a few terminals in the expectation of building on these later.

There was little evidence to suggest that the dichotomy was due to differences in the style of implementation. Two authorities, which approximated to a classically corporate style, had more than 16 terminals that could access GIS facilities, as had 2 of the theoretically/pragmatically corporate case studies. Although there was a link between departmental systems and fewer terminals, it is striking that over 50 per cent of the total number of terminals in the case studies were found in the eight planning and combined technical service departments involved in the various GIS projects examined. Even at a departmental scale it appeared necessary to have several terminals for the facilities to be widely available. It therefore seems that even among these early adopters, access to GIS capabilities was highly restricted. Trends will need to be monitored, but these indications suggest GIS to be regarded as relatively specialist technology (see also Payne 1993). Moreover, increased data sharing will be difficult if the information is not physically accessible.

The findings presented in Table 6.4 with respect to the types of departments that have computer access to GIS facilities confirms the significant involvement of technical service-type activities which was identified by the comprehensive survey. Not surprisingly, it was those departments whose responsibilities focus on the management of land and property including roads, the analysis of spatial trends and the provision of Ordnance Survey maps that showed the greatest

GIS and organizations

Table 6.4 Types of departments with computer access to GIS facilities

Types of department	Number with computer access to GIS
Planning/development	6
Information technology/computing	6
Engineers/surveyors/cleansing/highways	6
Environmental health	3
Combined technical services	2
Property/estates	2
Parks	2
Building control	2
Legal/land survey	2
Education	2
Architecture	1
Housing	1
Environmental health and housing	1
Planning and estates	1

interest in GIS technologies. Again planning departments are prominent. In contrast, even in these early adopters of GIS technology the large community service-type departments such as housing, education and social services had not perceived the benefits to be sufficient to become involved in these relatively well-established projects.

The discussion has so far concentrated on software and hardware. However, there are a number of important peripherals associated with the purchase of GIS. These include plotters and equipment such as digitizers. At the time of the fieldwork all but one of the case studies had acquired at least one plotter and all but three a digitizer. In the cases where automated mapping was the main focus of GIS activity, the availability and quality of plotters was as important as terminals. The case study findings indicate that while 36 departments had access to a terminal with GIS software, only 19 possessed some form of plotter. Furthermore the quality of the hard copy varied considerably. Most of the case studies had a range of plotting facilities including, at one extreme, basic screen dump printers through to AO pen plotters, and in two cases an electrostatic plotter. Access to better quality plotters was not limited to the more corporate developments as an electrostatic plotter had been acquired as part of a departmental system. In the non-departmental cases there was a tendency for one department, usually the section responsible for Ordnance Survey mapping, to have direct access to and therefore control over the best plotting facilities.

The findings indicate that the technical component of the GIS found in the case studies was very varied. The results also raise important questions about the physical accessibility of GIS facilities even in relatively well-established environments.

The human component

A critical element of any computer-based system is people, both in terms of their propensity to facilitate implementation and the cost implications of devoting staff time to a project. This section seeks to identify the staff resources and skills employed by the various case studies in the process of system implementation.

In all the case studies the staffing costs of developing a GIS were considerable and remained at a high level over a long period of time. There was some evidence to suggest that the levels of staff directly involved with supporting GIS technology increase over time as new applications are developed and the existing systems require ongoing maintenance. The majority of respondents had found that the staffing levels proposed at the start of system implementation had significantly underestimated the resources that subsequently had been deployed. As a result many were forced to gain approval from senior management and elected members for fresh injections of staff. In other cases changing priorities and reduced budgets had either prevented new staff being allocated to the project or squeezed existing resources despite requests to the contrary.

There are four types of basic skills required to implement a GIS. These are management capabilities, computing expertise, knowledge of data input and, depending on the applications being developed, specialist mapping and plotting skills. The rest of this section describes the human resources that the case studies were devoting to the various GIS projects. The impact of the differing approaches adopted to achieving effective implementation will be considered in the next chapter.

Table 6.5 provides an overview of the scale of staff resources that the case studies were expending on GIS development at the time of the fieldwork. There were considerable variations between the case studies, with a third having no full-time support while two-fifths had three or four staff working entirely on GIS-related activities. Table 6.5 indicates that the majority of full-time support for GIS development came in the form of specialist computing skills rather than the equally and perhaps even more important activities of data preparation and system management. The main responsibility for the implementation of GIS technologies in most cases rested with individuals who were very often being expected to graft these tasks onto their existing work-load. Only three case studies had less than four staff involved on a part-time basis in GIS development; two of those were departmental systems. In contrast, four of the case studies had at least three full-time and five part-time staff working on their respective GIS projects. Restrictions on finance often meant that staff resources were less than would be ideal. In a number of authorities, however, the low level of support also provided an indication of the relative priority given to the project.

One solution, which three of the case studies had adopted in order to supplement their limited staff resources and expertise, had been to contract out the technical side of system development and/or the time-consuming activity of data collection and input to a specialist company. For instance, one of the organizations with the most limited experience of spatial data processing had

GIS and organizations

Table 6.5 Range of skills and number of staff employed in implementing GIS technology

Skills	Full-time	Part-time	Total number of staff involved
Computer specialists	11	9.5	20.5
Data input	5	20.5	25.5
Plotting/mapping	2	2.0	4.0
Management/system coordination	4	15.0	19.0
TOTAL	22	47.0	69.0

Note: Part-timers were counted as 0.5

employed a private sector company to develop their entire GIS-based land charge searches system. Once completed, the company itself will receive ongoing revenue from the project, as a proportion of the fee for each transaction will be paid to them. In another case the authority had the necessary skills but lacked the staff resources to enable the input of the historic data for their development control system. This task was therefore given to a private contractor. In the final case contractors were employed to undertake the data collection. As a result in each of these projects the number of in-house staff involved in GIS development was far fewer than the average. It is, however, important to remember that if a computer-based system is to be effectively utilized there are ongoing costs in terms of maintenance as well as system modifications which will probably need to be met from internal resources.

For those case studies relying largely on in-house resources, the majority found that far more staff time had been required than was expected even to attain one operational application. The technological aspects of system development had been the area for which most full-time staff had been employed. At least one full-time computing specialist was working on half the systems. Like the technological determinist and managerial rationalist perspectives, there seemed to be an underlying assumption that system implementation is a largely technical activity. However, even given this general emphasis, two out of the three departmental systems and three of the theoretically/pragmatically corporate environments including the two with a long experience of geographic information handling were utilizing computing skills from within user departments rather than relying on central resources. In each of these cases it was felt that computing specialists in user departments had more experience of spatial data handling and a greater understanding of user needs than their colleagues in the central computing departments. However, this option was not available in most of the smaller authorities as they had no decentralized specialist computing staff. The computing personnel assigned to the GIS projects studied generally had no previous experience of GIS technology. As a result these skills tended to be acquired alongside the development of the GIS.

Table 6.5 indicates that while the greatest proportion of full-time staff working on GIS projects were computer specialists, there was a large group of individuals involved on a part-time basis with data preparation and input. Many of the respondents in the case studies emphasized that data collection, input and editing had proved to be a highly staff-intensive activity. The form of existing data sets had a strong bearing on the scale of the task. For those with partial data sets held on paper or no data at all, data collection rather than system development was the main focus of attention. The scale of this task is illustrated well by the highways management system being developed in one of the case studies. This application alone entailed the collection of five million separate pieces of information. Most of this work, including the subsequent process of inputting the data, was carried out by technicians and administrative staff from within user departments, with professional staff devoting considerable amounts of time to checking the accuracy of the information. As a result the development of a GIS application requires a significant commitment in terms of staff time on the part of user departments in a situation where the return on the investment is unlikely to be immediate.

Existing data sets in the majority of case studies in no way matched the standards necessary for a worthwhile GIS application. Many were incomplete or lacked any spatial reference, while for systems involving several data sets the absence of a common referencing system throughout the authority made modifications necessary. Only one authority already had a fully spatially referenced computer-based system prior to the start of the GIS project. In many cases data collection was viewed as a one-off activity with little consideration given to the establishment of a framework for updating the information, let alone the allocation of resources for this task. This is particularly important as the group that can perform this task most effectively often have the greatest pressure on their time, namely users. The discussion of data-related matters has so far concentrated on attribute data; however, four of the case studies had allocated staff time to handling the Ordnance Survey's digital map data. In these cases it was felt that coordination of the use of maps and the provision of plots required specialist skills, with one employing a cartographer.

These findings emphasize the extent to which data capture represents a highly significant element of the basic costs involved in developing a GIS. Initial budgets, however, often focused on equipment costs and as a result several of the case studies had found it necessary to seek additional funding for data collection. It was estimated in relation to one of the relatively small departmental systems that the data costs had been 16 times greater than the expenditure on equipment and support such as training. While this may be an extreme example, data capture was typically regarded as accounting for 70–80 per cent of total system costs.

The final set of skills involved in the development of a GIS is management expertise. Managerial skills of different types were stressed by both the managerial rationalist and social interactionist perspectives as having a significant influence on the effective implementation of computer-based systems.

The case study findings indicate that managerial input was mostly a part-time activity. Frequently, the coordination of a GIS project was added to a professional member of staff's already heavy work-load and therefore the time that could be devoted to GIS implementation was limited. Five of the authorities had appointed GIS project managers, often in desperation at the slow pace of development with only two of these working full-time on GIS matters. In a number of cases the individuals who were responsible for GIS developments devoted only part of their time to such matters and had someone working full-time under them. Four of the five organizations had located the individual concerned in the central computing section or the chief executive's department.

The case study findings suggest that the development of a GIS generally requires a substantial commitment of staff time. There can be no firm rules as to the number of staff or the types of skills required to implement a GIS, as much depends on the availability and format of existing data. In addition, the demands on staff time vary according to the type and scale of the application being developed. For instance, the level of support required to implement a relatively simple application such as automated mapping is less than complex applications involving the integration of several data sets. Overall, the balance of staff skills that the case studies had committed to the respective GIS projects reflects the widespread perception that the implementation of such systems is essentially a technical process. Implicit therefore is the assumption that if the technology will work, the system will be used. The extent to which this characterization of implementation is appropriate or helpful will be examined in the next chapter.

Organizational characteristics

This section focuses on the organizational characteristics of the GIS being implemented in the case studies. The managerial rationalist approach suggests that a corporate style of implementation based on rationally formulated procedures yields the greatest benefits. In contrast the social interactionist perspective places an emphasis on participatory and user-centred techniques, while such issues are of little concern to the technological determinist approach. The following discussion describes the circumstances found in the case studies in terms of the types and range of organizational units involved in the implementation process. It therefore focuses on the extent to which an essentially rational style of approach had been adopted.

Table 6.6 indicates that nine of the case studies involved at least two departments and therefore in some sense can be described as corporate, while a further three were developing independent departmental systems. The findings presented in this table highlight the strong contribution made by planning-related activities as the lead department for five out of the nine corporate systems was responsible for land use planning. In the other four, either there was no lead or GIS activities were coordinated by the central computing department. The link between planning and GIS is in line with the survey findings and reflects both

Reinvention and utilization

Table 6.6 Organizational structure for GIS implementation

Style of implementation	Corporate (lead department)	Departmental (department concerned)
Classically corporate	X (IT services)	
Classically corporate	X (Planning)	
Classically corporate	X (IT services)	
Classically corporate	X (None)	
Theoretically corporate	X (Development and technical services)	
Theoretically corporate	X (Planning)	
Theoretically corporate	X (None)	
Pragmatically corporate	X (Planning)	
Pragmatically corporate	X (Planning and estates)	
Fiercely independent		X (Parks)
Fiercely independent		X (Education)
Fiercely independent		X (Highways)
TOTAL	**9**	**3**

this department's traditional responsibility for Ordnance Survey mapping in most authorities and also their concern with the spatial distribution of activities. In terms of departmental developments, the involvement of activities such as highways and parks follows the national trends while the participation of an education department in the development of a GIS project was unusual.

In two of the organizations studied, corporate and departmental GIS were being simultaneously developed. In both cases the departments concerned had made positive decisions not to become involved with existing projects because of the immediate pressures created by the need to prepare for compulsory competitive tendering (CCT). However, the propensity for circumstances to change is exemplified by subsequent developments in one of the organizations.

GIS and organizations

Table 6.7 Number of departments involved and with access to GIS facilities

No. of departments	No. of departments involved with the project (no. of case studies)	No. of departments with access to GIS facilities (no. of case studies)
1	3	4
2	–	1
3	3	3
4	1	2
5	1	1
6	–	–
7	1	1
ALL	3	–

In this case the attempt to develop a separate departmental system was abandoned, largely due to encountering a wide range of technical and organizational problems, which were further exacerbated by the instability associated with the simultaneous reorganization of the department. Despite this experience, another department was granted the necessary resources to purchase its own GIS software, again as part of an attempt to enhance the competitiveness of a department facing CCT. The processes underlying these decisions will be explored more fully in the next chapter.

The findings presented in Table 6.7 indicate the number of departments involved with the development of GIS capabilities in the case studies. In terms of the joint systems, the number of departments participating in the respective GIS projects generally increased as the size of the authority decreased. As a result, the four smallest authorities attempted to devise structures including all or at least a large proportion of departments. In contrast, the other projects only involved three or four departments, with the exception of one which included seven. The practical limitations of achieving widespread participation are reflected in the second column of the table. This shows that while a certain number of departments may be regarded as involved in GIS implementation, in most cases the number that actually had direct access to the GIS software was far more restricted.

The review of the organizational structures that had evolved within each of the case studies has concentrated very much on internal arrangements. This is because there was no evidence in any of the authorities of inter-agency collaboration in GIS development. Moreover, in the vast majority of cases the software purchased was incompatible with the respective higher- or lower-tier authorities in the area. A similar situation applied to relations with the local utilities. General evidence suggests that the case studies were not unusual in this respect (see, for example, Onsrud and Rushton 1995).

Notions of corporate working as envisaged by the managerial rationalist perspective appear to be rarely applied in practice. At the most basic level the survey findings showed that the generally held assumptions behind the use of the term 'corporate', namely that all departments are active participants led by one of the central coordinating departments such as the chief executive's, are seldom realized in practice. The more detailed case study findings confirm this view and indicate the need to refine the initial typology of styles of implementation. As a result the three-level typology has been expanded into five stereotypes. The terms adopted for the revised typology draw on analogies with the family and highlight variations in the closeness of relationships between the organizational units participating in the GIS projects. The more detailed findings suggest that the representation of the organizational structure for GIS implementation in the formal documentation of the case studies only provides a stylized approximation to practice. Relationships within organizations, like families, seldom conform to an idealized model as well as being constantly under review. The revised stereotypes therefore take account of a number of additional factors including the ease with which informal networks between departments can develop, the location of the individuals with the influential role of establishing standards for spatial data handling, the extent to which user departments were responsible for funding and developing applications and the impact of external pressures such as central government's policies on privatization. The five stereotypes are as follows:

1. *Paternalistic* This refers to situations where there is a small staff and, as a result, a high level of personal contact between individuals in different departments. The organization may be referred to as a 'family' in which there are at times disagreements between members but on the whole there are strong personal ties to the authority. In these circumstances the expertise of the computing specialists in the central information technology department is widely accepted and they are given responsibility for directing the development of GIS facilities.

2. *Conforming adolescent* This stereotype refers to slightly larger organizations in which user departments take some responsibility for funding and developing their own GIS applications but within a framework set by the central information technology department. Experience of computing among user departments is generally very limited and therefore such matters are left to the technological specialists.

3. *Rebel adolescent* This term describes organizations that generally have large administrations including a number of key user departments with their own well-established centres of computing expertise. In this environment, standards for spatial data handling have already been formulated by at least one and possibly several user departments based on their long experience of processing geographic information. These departments tend to be regarded as mavericks, particularly by the central computing section. The participation of several

departments in a project such as the development of a GIS results from the formation of strategic alliances. Each department funds its own applications and there is always a chance that if the costs of working together appear to outweigh the benefits, a department will decide to go it alone.

4. *Newly independent* This refers to environments in which there has been a change from one of the above situations to circumstances in which GIS applications are developed on an essentially departmental basis. In these instances formerly conforming adolescent or even paternalistic environments have become highly fragmented as the authority has embraced a cost-centre-based administrative structure. Departments have therefore used their newfound independence to break away from the restrictions that had been an implicit part of a centrally coordinated and funded project.

5. *Fiercely independent* As in the initial typology this refers to the introduction and development of GIS technology by a single department. The impetus to act independently may emanate from a single individual or may be inherent to the department. Such departments have the facilities necessary to be largely self-sufficient in terms of the resources required to develop a GIS.

One issue which is important to bear in mind when considering this typology is the vast difference in the size of administrations investigated. For instance, the complete staff of one of the paternalistic environments amounted to 110, while the fiercely independent highways department had a staff of well over 500. Similarly, the conforming adolescent authorities had staff ranging from 400 to 600 while a single department in one of the rebel adolescent contexts employed nearly 1,400 people. Overall, the larger the authority the less closely different organizational units worked together. This emphasizes the importance of informal networks and familiarity with colleagues on the willingness of individuals to cooperate. Given the variations in size, proximity, either physical or psychological, is as important within departments as it appears to be between departments.

These findings confirm the earlier survey analysis, presented in Chapter 5, which suggested that few organizations in the local government sector have adopted a classically corporate style of GIS implementation. The extent to which this is a reflection of managerial inadequacies as the rationalist explanations would suggest or the inappropriateness of such an approach to the majority of the organizational cultures studied will be examined in the next chapter. However, these findings highlight the link between the implementation of an innovation such as GIS and existing organizational tensions. The issue of the distribution of responsibilities between the centre and individual departments is particularly important in this respect as it has a significant influence on the balance of power and control. This also takes a more specific form in terms of the relations between the centrally based computing specialists and user departments. In addition it is evident that seemingly unconnected events which alter relationships within an organization may have a profound impact on the process of GIS implementation.

Data

The key component of the basic fabric of any GIS is data. The types of geographic information held in such systems broadly take one of two forms. These are, first, digital map data which in Great Britain are largely provided by the Ordnance Survey, and second, attribute data which are largely generated by the departments themselves. At the time of the fieldwork, agreement between the Ordnance Survey and the local authority sector over the availability and price of maps was still being negotiated and consequently each authority had to make a separate purchase. Of the 12 case studies, 8 had acquired at least partial coverage for their area at the appropriate 1,250 or 2,500 scales from the Ordnance Survey. Five of these 8 authorities had complete coverage, while the remaining 3, including 2 counties with responsibilities for geographically large areas, had embarked on an ongoing programme of acquisition which would eventually result in complete coverage. Although the vast majority of the case studies were using vector-based maps, 2 had opted for scanned raster maps as a cheaper alternative which, it was felt, met their needs. This group consisted of one of the smaller authorities which was developing a land charges system and one of the departmental GIS. Neither of the 2 remaining departmental case studies had obtained any digital map data. In addition to the basic map base, 2 authorities had acquired the OSCAR (Ordnance Survey Centre Alignments of Roads) road

Table 6.8 Departments with attribute data on a GIS

Type of department	No. of departments with attribute data on a GIS in the case studies
Planning	8
Surveyors/engineers/highways	8
Environmental health	4
Parks/recreation/amenities	3
Legal/land charges	3
Property/estates	2
Housing	2
Cleansing	1
Planning and estates	1
Education	1
Environmental health and housing	1
Building control	1
Building and works	1

GIS and organizations

network data for their areas as the backcloth for census mapping and highways management systems.

Table 6.8 indicates the range of departments that had attribute data on the various systems being developed in the case studies. This is perhaps one of the best indicators of the types of departments that had made a commitment to the development of GIS technology, as the collection and input of attribute data was largely the responsibility of individual user departments. In some cases this was a purely technical operation involving the transfer of data sets between different computer software, but for most the task was much more staff-intensive and lengthy. Many of the data sets were in a partial form and, due to changes in priorities, it appeared likely that several would never be completed.

In line with the earlier findings the technical service departments were the most likely to have input data into a GIS. The planning and surveying functions showed the most pronounced commitment, accounting for around 44 per cent of all the departments with data on a GIS. However, nearly a quarter of the departments listed in Table 6.8 are accounted for by a single case study. The next section examines the types of applications for which these data sets were being used.

Applications

Tables 6.9 and 6.10 indicate the types of GIS applications that were being developed in the case studies. Table 6.9 identifies those applications that were technically operational at the time of the fieldwork, while Table 6.10 points to the applications that were still under development. More recent information suggests that it is unlikely that more than one or two of the applications outlined in Table 6.10 have since been completed despite technical trials having been undertaken for a small geographical area. These findings indicate that some departments that had input attribute data into a system had not gone on to develop an application. For example, four environmental health departments had data sets on a GIS but no applications had resulted. This is because these often small data sets were input on an experimental basis, often by another department and with no specific objective.

The range of applications listed in Tables 6.9 and 6.10 suggests that the case studies were concentrating on utilizing GIS technology to assist with operational activities such as grounds maintenance, highways management and planning applications processing. However, the greatest current and anticipated use of the technology was for the production of basic maps. This suggests that, at least for the foreseeable future, GIS technology will impact most strongly on the work of technicians and administrative staff rather than professional and managerial levels. The development of strategic decision support systems was not regarded as a priority, even in the authorities where some form of corporate approach had been adopted. A number of respondents who had had experience of developing corporate databases during the 1960s and 1970s stressed that they perceived the

Table 6.9 Operational applications

Application	No. of users*	Functionality
OS map production (1)	22	Display
OS map production (2)	17	Display
Council house sales	11	Display
Vacant, housing/industrial land analysis	7	Display Query Analysis
Forward planning analysis	3	Display Query Analysis
Grounds maintenance for schools	1	Display Query
Planning applications processing	0	Display Query
Production of the planning register	0	Display
CCT contract for cleansing	0	Display Query

*These figures are indicative of the number of individuals making direct use of the applications at the time of the fieldwork

Table 6.10 Applications under development

Applications	No. of case studies
OS map production	7
Highways management	3
General database development	2
Planning applications processing	2
Grounds maintenance	2
Analysis of land use data	2
Environmental protection legislation – litter zones	1
Property management	1
Land and property management	1
Land charges	1
Council house sales	1
Tendering document for cleansing	1

main value of computer-based systems such as GIS to lie in simple repetitive tasks. These discussions did not suggest that those in practice expected computer applications to follow an inevitable progression starting with simple operational tasks and moving through to strategic decision-making.

One of the main criteria for assessing the effectiveness of the implementation process is whether use is being made of the applications. For the technological determinist and managerial rationalist perspectives, utilization is regarded as inevitable given that the system is technically operational and a rational managerial strategy has been adopted. The social interactionist perceptive, in contrast, considers the process to be far more complex with no certainty that a technically operational system will be used. The following discussion examines the use being made of GIS capabilities in the case studies, with the term 'user' referring to individuals who interact directly with the technology. This operational definition leads to the omission of those who might make considerable use of the information generated but do not come into contact with the GIS facilities. It was felt that at this stage in the analysis the basic information provided an adequate indication of trends.

The findings presented in Tables 6.9 and 6.10 are striking in two respects. First, despite the presence of GIS facilities within the organizations for at least two years, very few applications were technically operational. Second, only a limited number of these operational applications were actually being used. These findings are summarized in Table 6.11. It was possible in only three case studies to identify any users of the GIS technology adopted, while a further two had developed technically operational applications but the systems were not being utilized by users. In the latter case, potential users could not envisage any circumstances in which they would make use of the particular application. These systems were classified in the middle category alongside those that were still under development. In a further two projects the work on applications had been abandoned and the future of GIS within the organization was being reviewed. One of these was a departmental initiative while the other involved several departments in what was envisaged as a major authority-wide project.

The three case studies that had achieved utilization of their GIS facilities had adopted very different approaches. These were a small, essentially one-person departmental system, a relatively simple system with an average number of support staff fitting the conforming adolescent stereotype and a large-scale project approximating to the rebel adolescent description. None of the smaller authorities had as yet any users of the applications they had been developing. In fact the large-scale system accounted for three of the six applications and nearly two-thirds of the users. Despite the multi-departmental nature of the arrangements for GIS development in two out of the three cases, all the applications, with the exception of the production of Ordnance Survey maps, were based within individual departments and served their needs alone. The situation with respect to the applications still under development was less pronounced, with slightly under half based exclusively within a single department. It is possible that the involvement of a greater number of organizational units accounts for the

Table 6.11 Overview of the utilization of GIS in the case studies

Result	No. of case studies
At least one application operational with one user	3
Still developing the system or an application operational but not being used	7
System abandoned, review of the future of GIS under way	2

slower pace of development of some of these systems.

A common feature underlying the three seemingly dissimilar and even contradictory environments that had achieved utilization was the simplicity of central application which they had developed. Two had concentrated on automating the production of maps while the departmental system focused on assisting with the tendering procedures for the grounds maintenance of the county's schools. With respect to automated mapping, the production of Ordnance Survey maps by non-specialist staff was a relatively common occurrence in three departments in one authority and four in the other. In both cases this highly visible initiative had given the project credibility and thereby encouraged the development of further applications.

For the case studies that were yet to complete an application, the introduction of automated mapping facilities was usually seen as secondary to the development of an application involving a major data capture exercise. In such circumstances it is likely that commitment to the project will have to be sustained over a considerable period of time during which the return is limited. The findings confirm this, indicating that the time that had elapsed since the purchase of GIS software varied from two and a half years to seven, with an average of around four years. The extremes for the case studies with applications that were being utilized was not as great, ranging from one and a half years to four years. These differences reflect the scale of the projects, the quality and availability of the existing data, the skill with which the system was implemented as well as some measure of good fortune. The underlying processes will be explored in the next chapter; however, these findings emphasize the length of time that it can take to realize benefits from a technology such as GIS. This is an important consideration in circumstances where there is a demand for short-term gains on investment.

One of the key debates in the GIS literature has been the type and level of functionality that should be included in such software. Inherent within much of this discussion has been the technological determinist assumption that greater technical sophistication leads to adoption and utilization. Table 6.9 provides an indication of the types of facilities that were being required of the applications operational at the time of the fieldwork. The applications that were in greatest

GIS and organizations

demand were those that facilitated the display of maps, such as up-to-date Ordnance Survey maps or the property boundaries of council houses. The second most important group of facilities that GIS were expected to provide were query and search capabilities. Analytical facilities were regarded as having a relatively low priority in all the case studies. The only exception to this were a few individuals based in planning departments. Table 6.9 indicates that two of the planning departments in the case studies were analysing changing land use patterns using GIS capabilities. In each of these cases the analytical facilities being used were relatively simple, such as calculating areas and overlaying data sets. Even in these relatively advanced environments there was no demand for modelling capabilities. Moreover none of the case studies regarded the current level of functionality of the GIS they had purchased as in any way inhibiting system development. The main demand was for improvements to the ease with which simple tasks such as displaying and plotting maps could be undertaken, rather than the provision of additional analytical capabilities.

These findings suggest implementation to be a problematic and uncertain process. There is no guarantee, as the technological determinist and managerial rationalist perspectives tend to assume, that the logic behind the decision to adopt a technological innovation will be sufficient to ensure its utilization. In all cases implementation of just one application was a lengthy process while several cases demonstrated that technical success was not sufficient to achieve utilization. There was some indication that the simpler the application, the more likely it was to be effectively implemented, although a few organizations had managed to achieve the utilization of relatively complex systems. This raises important issues about the nature of implementation which will be examined in the next chapter.

Evaluation

The preceding discussion has explored the reasons behind the decisions of the case study organizations to adopt GIS technologies, described the technical and organizational characteristics of the systems being implemented and examined the extent to which the facilities were being used. The processes underlying implementation will form the focus of the next chapter but it is important at this stage to consider two issues raised by the theoretical discussion, namely reinvention and the nature of success and failure.

Reinvention

The findings of the case studies show that technological innovations are not discrete entities consisting simply of a particular configuration of equipment. The vast majority of those interviewed related to GIS technologies in terms of its social meaning for them as individuals rather than its technical components.

Such circumstances are common as it is not necessary to understand the workings of the internal combustion engine to purchase a car. The acquisition of an innovation is largely therefore an act of faith, but the beliefs that prompt the purchase appear to vary considerably. It is a particular feature of programmable technologies that they offer the capacity to be manipulated into a variety of different forms, depending on the needs of individuals and the organizational context. In many ways it is precisely this capacity for radical and ongoing reinvention that distinguishes these technologies.

The tangible impacts of the process of reinvention can be seen in the basic components of the GIS being introduced in the case studies. Each had chosen a different configuration of equipment, filled it with different data and wrapped it up in different organizational structures. This outcome is the result of a series of personal responses set within a particular organizational setting. At the organizational scale technologies such as GIS are largely viewed in terms of specific applications. Academic discussions as to exactly what functional capabilities a technology must have to be classed as a GIS are largely an irrelevance to potential users. The technology is not perceived in the raw, rather it is seen as a system to assist with the reproduction of maps or management of highways, a tool box able to contribute to a multiplicity of activities or even, by some, a plant pot stand. It is unlikely that two GIS-based highways management systems will be the same as each organization explores and then reinvents the technology to fit the particular context. The process of reinvention therefore continues until the technology and the organizational context reach some level of equilibrium. If the technology is regarded as alien to the culture it is likely to be abandoned.

Alongside the physical development of the technology, individuals within an organization make their own decisions as to its social and political implications. Each individual will attempt to maximize their personal gains and minimize their losses. The initial discussion concerning the reasons for GIS adoption in the case studies highlights the social and symbolic aspects of innovations. As a result, in addition to a GIS gaining meaning as an application it also has a very important symbolic function. In some cases an individual may attempt to reinvent an organization in the image of an innovation, although it would seem probable that in the end organizations reinvent new technologies in their own image. Several of the case studies suggested that their understanding of GIS technology had changed during the course of system development. Initial expectations were shaped by vendors but as individuals within the organization interacted with the technology, their perceptions changed. This was seen as an ongoing process as experience grew and circumstances changed.

The research has by no means fully explored the concept and nature of reinvention. However, the findings suggest that the meaning of a technology such as GIS was continuously being reinvented at both the organizational and individual scales. This in turn has important implications for studies of diffusion as it would appear that innovations such as GIS embrace a wide range of perceptions. These differences in emphasis are likely to lead to tensions and problems which will complicate the implementation process. It is also likely that such systems will be used to undertake activities not originally anticipated by their inventors.

Success and failure in GIS implementation

There is often a one-dimensional quality about the use of the terms 'success' and 'failure' in relation to the implementation of information technologies. Implicit is the assumption that a computer-based system is either a success or a failure. Such assessments are often made without identifying the criteria on which the judgement is based. This tends to result in a rather mechanistic approach to the process with the key questions centring on such issues as: what is the best software? How many staff are required to ensure success? What organizational structures should be adopted in order to prevent failure? However, success and failure rarely appears to be as clear cut as this. The processes that affect the effective implementation of GIS technologies will be considered in the next chapter, but first it is important to consider the basic nature of success and failure in relation to the case study findings. The typology of perspectives on failure discussed in Chapter 3 provides the framework.

An important issue to bear in mind before examining the experiences of the case studies is the extent to which judgements as to the outcome of implementation change over time. The dynamic nature of the environment in which any computer-based system is embedded makes assessments as to the relative balance between the successful and less successful aspects of a project time-specific. Regardless of the criteria that are chosen, success in the short term may become failure in the medium to long term. Equally an early false start does not necessarily imply that the project is destined to fail. Evidence from ongoing contacts with the case studies suggests that both circumstances may arise. The following comments therefore draw on the findings derived from the longitudinal aspects of the research.

One approach to success and failure is to examine the results of implementation in terms of the objectives set out at the start of the project. Lyytinen and Hirschheim (1987) term this 'correspondence failure'. The case study findings demonstrate such an assessment to be difficult and in many cases misleading. It assumes that the adoption of an innovation is based on clearly identifiable goals and that formal expressions of objectives reflect underlying motivations. In some of the case studies, particularly those in which a GIS was purchased as part of a large investment in information technology, it was difficult to identify specific goals. Moreover, where the stimulus for adoption was symbolic and political, formal statements of intent are likely to be at variance with the underlying agenda. For instance, where a new technology such as GIS was purchased to mark a change of administrative approach, it may not matter whether the system is ever used. In this case the critical part of the process is the initial action. In contrast, where a department is introducing a GIS to signal its self-sufficiency, the visibility of the facilities and the information generated is likely to be very important. Consequently, a system may fail to meet its stated objectives yet be regarded as a tactical and political success. At a more practical level, in several of the case studies circumstances had changed to such an extent that the initial objectives were no longer appropriate. Mechanistic application of correspondence as the basis for judging success and failure would therefore appear to be of limited value.

Relative levels of success in the implementation of computer-based systems can also be assessed on the basis of the process. Lyytinen and Hirschheim use the term 'process failure' to refer to situations where an operational system cannot be achieved within the initial budget guidelines. Utilizing this criterion, it would seem that none of the case studies could be regarded as a success. At some point in the implementation process all had to seek additional funding. Process failure, like correspondence failure, therefore appears to assume that all aspects of the process can be known at the start and that circumstances will remain constant. Experience in practice casts considerable doubt on the merits of such criteria as they are often difficult to apply and provide limited insight.

A further dimension of success is whether the system fulfils expectations. This approach takes account of implicit as well as explicit objectives, while enabling perceptions to be coloured by experience of the process. The main difficulty in applying this criterion is whose views should take precedence. Organizations are not single entities, they consist of departments, interest groups and individuals, each with their own priorities and concerns. At the level of the individual one person's interesting computing problem or good career move is another's administrative inconvenience or lost job. It was therefore difficult to obtain consensus in many of the case study organizations as to whether the GIS they were implementing was fulfilling expectations. Many of those not directly involved with the decision to adopt a GIS emphasized that they had no expectations for the system. As a result, measuring success in terms of the realization of expectations adds to the overall picture and often highlights conflicts of interest and values but is difficult to apply in practice.

The final criterion that Lyytinen and Hirschheim identified was interaction failure. This focuses on the extent to which an innovation is used. The main limitation of this approach is where users have no option but to employ the new system in their work. In such circumstances use is necessarily a poor measure of satisfaction. However, in none of the case studies was the use of a GIS the only way of undertaking a particular task. Consequently, the basic action of utilizing the system provides at least an indication of user satisfaction. The extent and frequency of this interaction provides a further guide as to the perceived value of the innovation. However, it is important to bear in mind that some groups and individuals within an organization will not share this perspective on the system. Moreover, as the case studies demonstrated, it is quite possible that a system is regarded as a technological or symbolic success yet is not used.

The findings of the case studies indicate the complexity of applying the terms 'success' and 'failure' to the implementation of technological innovations such as GIS. All systems combine some elements of success and failure with the precise balance varying over time. It is striking that, regardless of the criteria chosen, success had proved extremely elusive in the organizations studied. The subsequent analysis will examine why the implementation of GIS technologies appears to be so problematic. In particular the discussion will explore the issues surrounding the limited utilization of such systems as the basic concept of use appears to encompass most fully the various aspects of success and failure in relation to GIS. However, given the various dimensions to success, it will also be important to take account of the symbolic role of the technology and the differing perspectives of interest groups within organizations on the merits of the systems being implemented.

CHAPTER 7

GIS implementation: a problematic process

Introduction

The research findings suggest that realizing the theoretical potential of a technological innovation is a difficult process. The case studies demonstrate that the mere purchase of a GIS by no means ensures utilization. It is therefore important to explore the nature of implementation in more detail as it is this process that is responsible for transforming boxes of equipment into a taken-for-granted part of the operations of an organization. The theoretical framework devised in Chapter 3 provides two levels of insight into the implementation process. Each of the three perspectives suggests a particular prescription as to the most appropriate methods to adopt in order to achieve utilization. In addition, an explanatory framework is proposed through which difficulties in securing this objective can be understood. The key elements of the various perspectives are summarized in Table 7.1.

The theoretical framework suggests that there are three groups of explanations for the difficulties encountered by the case studies in effectively implementing GIS. These are basic inadequacies in the technical performance of the innovation, suboptimal managerial strategies and the interaction of social and political processes within a particular organizational environment. The following discussion therefore explores the extent to which these explanations appear to account for the experiences of the case studies and the degree to which the various prescriptions were followed. Consideration of the managerial and organizational aspects of implementation is structured around the key issues which were identified in Chapter 3 as facilitating implementation according to the social interactionist perspective. These are as follows:

1. an information strategy that identifies the needs of users and takes account

GIS and organizations

Table 7.1 Perspectives on the implementation process

Perspective	Prescription	Explanatory framework for poor performance
Technological determinism	Data-processing approach focusing on technological development	Inadequate technology
Managerial rationalism	Data-processing approach or Structured design methods, based on corporate working in order to facilitate data sharing	Inadequate management strategy
Social interactionism	User-centred design philosophies incorporating participative approaches, sensitivity to change and awareness of cultural uniqueness	Uncertain interplay of social and political processes

of the resources and values of the organization;
2. commitment to and participation in the implementation of the system by individuals at all levels of the organization;
3. an ability to cope with change.

This is a useful device for organizing the subsequent analysis as each embodies a set of assumptions that counter traditionally based rational conceptions. Therefore it will be simultaneously possible to examine the validity of the social interactionist prescription and also examine the assumptions underlying the managerial rationalist approach. In particular the discussion concerning the preparation of information management strategies will explore the extent to which data sharing was given priority by users while the propensity for corporate working will be considered in relation to the development of commitment and participation in the various GIS projects studied. However, before exploring the managerial and organizational aspects of implementation, the next section briefly considers the technological experiences of the case studies.

Technological considerations

The technological determinist perspective conceptualizes implementation as largely a matter of developing technically operational applications. Failure to achieve utilization is therefore regarded as a result of inadequacies in the technology or the incompetence of users. The skills and expertise of users will be examined later in the chapter; this section concentrates on assessing the technical worth of the GIS being implemented in the case studies. This issue is particularly important, given the timing of the research, as all the case studies may be regarded as pioneers in the GIS field. As such it is probable that both the software and commercially available data that they purchased were more temperamental and prone to error than later developments. The subsequent discussion will seek to examine whether the problematic nature of the implementation process in the case study organizations can be accounted for by the technical inadequacies of the GIS purchased. The findings have been divided into two parts, with the first section focusing on software and hardware issues and the second on deficiencies in the commercially available data which usually manifest themselves as technical problems. It should be emphasized that the objective of this discussion is not to present a technical evaluation of the merits of the various software packages adopted by the organizations studied. Rather, the aim of the analysis is to report on the perceptions of those interviewed and assess the extent to which the technological problems encountered influenced the implementation process.

Software and hardware limitations

It is widely accepted that the development of computer-based applications entails solving technical conundrums at the very least and possibly more profound technological problems. The particular requirements of each organization inevitably pose new demands for the available software, some of which will have been anticipated while others may stretch the technology to its limits. The experiences of the case studies with respect to GIS technologies confirm these general trends. Three-quarters of the case studies mentioned that they had encountered technological difficulties of one sort or another during implementation.

The issue of greatest concern was the level of customization necessary in order to allow even the most basic use of the system by those with advanced computing skills. Moreover, the next stage of making the software sufficiently user-friendly to enable non-specialists to utilize the system was regarded by many as difficult and extremely time-consuming. Since the research was conducted there have undoubtedly been important enhancements to the commercially available technologies including considerable developments in user interfaces. However, the very essence of programmable technologies is the flexibility they offer, enabling systems to be devised that meet the needs of particular organizational environments. The greater the flexibility in terms of the types of application that can be developed from a particular GIS software package, the longer it will take to achieve an operational system and the more extensive the customizing requirements are likely to be. The issue of customizing, perhaps more appropriately domesticating, a technology is not therefore symptomatic of the stage at which GIS software had reached at the time of the fieldwork. Moreover, nor is it exclusively associated with the development of GIS technologies as very similar issues arise in relation to management information systems, for example.

The extent of customization necessary in the various organizations studied depended on the type of applications demanded of the software adopted. Some systems and applications required considerably more development time and technical expertise than in other cases. However, the impact of these technological difficulties on the process of implementation appeared to have more to do with the response of the individuals within the organization than the scale of the problem encountered. There was a tendency for staff in the more experienced computing environments to have an expectation that extensive programming work would be necessary and therefore not to view it as a particular problem. Difficulties and setbacks in these circumstances were regarded as part of the process of undertaking a new activity for the organization rather than an issue of more profound significance. In these contexts the form of technology appeared to be largely irrelevant to the outcome of the implementation process. For example, one of the most heavily criticized GIS products was being made to work by one of the case study organizations. In contrast the response of some of the less experienced environments was more negative, provoking some to question the whole validity of their particular project.

The same variety of responses to other technological difficulties could be seen in the case studies. These problems included the failure of specific components of the software to operate, inability to link the GIS to existing software in the organization and inadequate data storage capabilities due to the nature of the underlying data model. Hardware issues were also a matter of concern for some of the case studies. While hardware is not a problem that exclusively confronts GIS developments, it is an integral part of such systems. The most common concern in this respect was the poor reliability and limited capabilities of the computing equipment, although overall improvements in this field, including vast increases in the computing capacity of micros, appeared to be enhancing flexibility rather than constraining GIS implementation.

In general the majority of the technological problems encountered were being resolved, given sufficient time and the necessary resources. This in turn raises questions about priorities and the ability of the organization to cope with the uncertainties of change. The implications of these findings will be considered after examining the technical limitations of the commercially available spatial data.

Limitations of the commercially available spatial data

The key commercial spatial data set for the case studies was the Ordnance Survey's digital maps. At the time of the fieldwork each digital map had to be purchased separately by local authorities. Moreover, the digital coverage of

Britain was not yet complete and those who had purchased data were concerned about their accuracy, up-to-dateness and whether the maps matched at the edges. The majority of these technical problems have since been overcome as the Service Level Agreement between the local government community and the Ordnance Survey bears witness. This agreement also tackles the complex issues of copyright and charging which had caused difficulties in the relations between the two groups.

The findings of the case studies suggest that the limitations of the digital data available at this time in no way accounted for the failure of users to utilize the GIS facilities within their organization. At their most profound, such problems merely appeared to reinforce existing frustration and disillusionment with a system that had been rejected by users for other reasons.

Evaluation

All the case studies encountered a mixture of anticipated and unforeseen technological problems during the implementation process. In terms of the software, there was no link between the type of product that had been acquired and whether utilization was achieved. The demands placed on some software products proved more problematic than for others but such issues did not appear to determine the outcome of the process. It was possible to identify a couple of instances where technical deficiencies in the systems being implemented contributed to the slow pace of development. However, other environments faced similar circumstances and yet the impact on progress was far less marked.

Technological developments in both computing facilities and commercially available data have undoubtedly improved the quality and ease of use of GIS although they have by no means removed the need for system development following acquisition. Even given the technological difficulties faced by the case studies, respondents generally regarded them as nuisance rather than an insurmountable problem. The issue was not whether the difficulty could be resolved but whether the project was of sufficient priority to the organization that the necessary resources and staff time would be allocated to the matter. Furthermore, some organizations were better able to cope with these problems and eventually overcome the hurdles than others. Some even demonstrated a capacity to turn such problems to their advantage; as one respondent put it, 'there's nothing like a crisis to get things moving'. Consequently, technological issues do not themselves account for the propensity for GIS to be utilized but rather the manner in which individuals within particular organizational environments respond to such problems. In some cases these issues reinforce existing doubts about the value of the innovation while in others such difficulties are viewed more as an intellectual puzzle which is an inevitable part of the implementation process. Moreover, as the findings emphasize, there is no certainty that technically operational systems will be utilized. This therefore casts doubt on the assumptions underlying technological determinist explanations. The implementation of innovations appears not simply to be a matter of technological advancement but to be a social and political process. As a result the remainder of the chapter will examine the managerial and organizational explanations which seek to account for the experiences of users.

Organizational considerations

The managerial rationalist and social interactionist perspectives view difficulties in securing effective implementation to be a consequence of either inadequacies in the managerial strategies adopted or a lack of accord between the technology and the values and practices of the particular organizational culture. The following discussion explores the merits of these explanatory frameworks in the light of the case study findings. The analysis is divided into three parts. The first examines the information management strategies adopted by the case studies, the second the level of commitment and participation in the various projects and the third the capacity of the organization to cope with change. In each case the managerial strategies adopted by the case studies will be described, followed by an examination of the underlying processes that appear to account for their experiences.

Information management strategies

The potential importance of information management strategies for the effective implementation of GIS was outlined in Chapter 3. The social interactionist perspective regards these strategies as embodying the information priorities of users based on available resources and the values and practices of the organization. The objective, therefore, is to embrace the organization as it is rather than impose an alien structure. There is no prescribed approach as to the form that such strategies should take: they may be informal or formal, written or unwritten, departmental or organization-wide. In contrast, the managerial rationalist perspective places an emphasis on strategies based on the data-processing model or structured design methods. In the case of GIS this has led to an emphasis being placed on data sharing as the underlying rationale.

The following discussion explores the role of information management strategies in achieving effective implementation in the case studies. The first part of the analysis examines the extent to which the case studies had prepared strategies and what form they took. As only very few had been produced, the second part investigates the potential content of such strategies. In particular the assumptions underlying the managerial rationalist perspective as to the needs of users and the resources at the disposal of the organizations to develop a technological innovation such as GIS are investigated. One of the most widely promoted justifications for GIS has been its ability to facilitate data sharing and thereby satisfy a supposedly latent demand among users.

Information management strategies and the case studies

Only two of the case studies had produced a written information management strategy, while a third was embarking on the process at the time of the fieldwork. The most comprehensive approach combined an authority-wide strategy with individual departmental documents which had to be consistent with the overall framework. In the second case an experienced information handling department which was part of a multi-user project had devised their own strategy. Given the central role of this department within the authority in terms of information processing, the document produced had implications beyond the confines of that particular department. The circumstances and likely outcome of the strategy currently under preparation were very similar to this with a single department taking the initiative. Both the completed strategies had received approval from elected members and were located in very similar types of environment. For instance, both had faced the destabilizing effects of the introduction of a cost-centre-based management structure.

The key question therefore is, what has been the impact of the framework provided by the information management strategy on the implementation of GIS in these two organizations? The findings of the case studies are by no means clear cut. In one case the likelihood of achieving utilization appeared to be highly probable while in the second it was unlikely. In contrast to managerial rationalist expectations, it was the context that had produced an authority-wide strategy that was experiencing the greatest problems. This suggests that the mere production of a strategy is insufficient to ensure effective implementation. A review of the strategies indicated that both the process and the final product had proved more valuable where the objective was to formulate a robust framework which identified the types of information that were essential to the operation and values of the organization rather than simply 'nice to have'. An important additional issue was the extent to which knowledge of the role and importance of the core data sets was shared by staff throughout the organization. These limited findings indicate that the production of an information management strategy on its own is insufficient to ensure the effective implementation of a GIS. However, it was evident that without an understanding of the information priorities of the organization and the contribution of technology to achieving these goals, the sustained utilization of GIS facilities appeared to be unlikely.

The limited number of information management strategies among the case studies contrasts with that of information technology (IT) strategies. The central computing section of every authority had devised an IT strategy in line with the advice given to local authorities by the Audit Commission (1990). The contribution of these documents to the development of GIS in the case studies appeared to be limited, with staff in many of the user departments unaware of the content of the strategies or else antagonistic to what they regarded as the centralizing tendencies of the underlying philosophy. It was also evident that the majority of the strategies were mostly concerned with financial systems rather than spatial data. Moreover, emphasis was on technological matters rather than the information priorities of the users.

These findings suggest that the general framework for GIS development in the case studies was largely provided by strategies that placed the main emphasis on technology rather than information or the organizational context. This pattern was reflected in the detailed programmes that had been produced to guide the specific implementation of GIS in the various organizations. Seven of the case studies had prepared GIS strategies, including one of the departmental systems. All had a tendency to favour the data-processing model and thereby to focus on technical matters, with two giving little consideration to anything else. Key characteristics of the remainder were that two were departmentally based, one was entirely informal, another was produced some time after the initial purchase of the system and the last consisted of the unwritten accepted wisdom of one individual. The last reflects the long experience and organizational standing of the member of staff concerned. Analysis of the content of these strategies indicates that the link between the potential information to be generated by the GIS and the essential needs and values of the organization at either an authority-wide or departmental level was very poorly developed. Most strategies were dominated by assumptions about user needs rather than actual attempts to identify their requirements and on this basis establish the most appropriate methods to supply the information required as well as the necessary skills to exploit any computer-based developments. It is interesting that perhaps the most effective of the GIS strategies examined was the one often referred to as a particular individual's 'head'. Such strategies reflected the widespread acceptance of this individual's expertise and organizational awareness. As a result these findings imply the importance of a shared understanding among all staff within an organization of the role and value of geographic information in their work rather than the need for reams of documentation.

In terms of the prescriptive recommendations, the approaches adopted by the majority of the case studies follow more of a managerial rationalist than a social interactionist philosophy. The overall focus tends to be placed on the technology with assumptions made about information needs and the context into which the system is to be located. It was evident from the interviews conducted with staff and senior management in prospective user departments that many had not been involved in identifying the goals of the various projects. As a result, given that the overall findings indicate a reluctance among some users to utilize the GIS being implemented, it is important to examine whether there is any mismatch between the often implicit assumptions as to user needs and their requirements in practice.

User needs

The process of defining user requirements is by no means straightforward. System designers often comment that users do not know what they want. Experience drawn from the implementation of computer-based systems in general shows that effective utilization is seldom achieved where computer specialists make assumptions about the needs of users (see, for example, Eason 1988). Even in the context of a departmental GIS project, a system developer in one of the case studies remarked that defining user needs was one of the hardest yet also most critical aspects of the implementation process.

It is therefore important to explore how users in the case study organizations conceptualized their needs with respect to GIS. The analysis considers user requirements in terms of not only geographic information, but also the type of service provided and the form of the technology. However, an important issue which underpins any discussion about GIS concerns the assumption that there is a latent demand among users to share geographically based data sets which can be satisfied through the ability of GIS to integrate spatial data from different sources. This issue will be examined first.

Data sharing

Many of the arguments justifying the adoption of GIS are based on the capacity of this technology to facilitate data sharing (Masser and Campbell 1994). These arguments presume the existence of what is often referred to as corporate information. This is assumed to be information that is of value to all or at least most of the separate units into which an organization is divided. Several studies have produced matrices indicating the widespread demand for particular spatial data sets by all departments within a single organization (Bromley and Selman 1992). Given that the existing facilities provide no mechanism for the general distribution of these data sets, it is logically presumed that GIS technology can enhance efficiency, improve strategic, managerial and operational decision-making and reduce wasteful duplication. The first issue to consider therefore is the extent to which there are corporate spatial data sets within the organizations studied and, second, whether it is merely technological constraints that are preventing the exchange of this information.

The experiences of one of the county councils studied were highly instructive with respect to the nature and existence of corporate data sets. This authority, spurred on by the appointment of a new chief executive, has implemented a corporate information network. The aim was for all data sets held by any unit within the organization to be made available through this system. A comprehensive network of terminals had been established, including distributed sites such as schools, so as to provide ready access to the information. This experience indicated that very little, if any, of the information available was accessed by more than the original holding department. Furthermore, as more and more data sets were being added to the system it was becoming increasingly difficult to locate the particular information required while response times were tending to slow down. Consequently, given the daily operational pressures on staff, few felt that they had the time or the inclination to search through the system for information that might be of potential interest.

These findings suggest that the only widely used spatial data set within the authority was Ordnance Survey maps. These sentiments were by no means limited to this one organization. Rather, they reflected the experiences of most of the case studies that had attempted some measure of authority-wide implementation. In some cases this realization had come earlier than others. One of the most experienced information handling environments had rejected the idea of corporate data in favour of what were termed 'functional clusters'. This approach was based on the notion that there were groups of departments or even sections that share common interests and for which it might be possible to achieve some form of joint working. Examples of the type of information that formed the basis of these clusters were legal, geotechnical and control data.

Even at the level of Ordnance Survey maps there was general evidence of individual departmental interests inhibiting joint working. For instance, those departments that had traditionally been heavy map users and had developed a large measure of independence over the provision of these needs were extremely wary of relinquishing this control to a central map unit. Furthermore, the mapping needs of different departments varied considerably. It was evident that while different sections may utilize the same basic data sets, what they want to do with them and therefore the form in which they require such information often differs significantly. Census data are another good example of this. Many of the departments in the case study organizations wanted to utilize this information but they tended to be interested in different elements of the data set, as well as wishing to analyse it in very different ways and at widely differing scales. Consequently, centralizing the provision of geographic information is likely to be extremely difficult with the possibility of no one's needs being satisfied. It would appear that functional clusters may prove a more effective means of exchanging data as it is likely that all those handling legal information, for example, would have similar needs and make similar assumptions about the nature of that information.

Overall the case study findings indicate that in practice very little data sharing took place, not only between departments but also between sections within the same department. This appeared to have little to do with technical constraints but rather the nature of information. Individuals within the organizations studied did not regard information as merely units of data. Nor was it perceived to be value-neutral, objective or self-explanatory. For instance, in seeking information, individuals frequently did not just want an item of data. More important was the associated professional opinion as to the most appropriate manner in which to interpret the data. As a result it was often felt to be more efficient to contact the appropriate member of staff in the first place rather than refer to the GIS and then request further details from the individual concerned. Many of those already familiar with a particular data set saw little need for on-line facilities as their familiarity caused them to have few queries. As a result they felt that they were unlikely to remember the commands necessary to make use of a computer-based system. Many of the respondents in this category demonstrated a clear preference for utilizing hard copy.

A further issue with respect to the nature of information was the widespread concern among many professional staff that there was a risk of misinterpretation if certain data sets were made available. Figures alone do not convey the customary practices that account for the level of accuracy with which a particular set of information should be treated. It was felt that computer-based systems tend to compound the chance of misunderstandings as they are often falsely associated with accuracy.

The introduction of compulsory competitive tendering and cost-centre-based management structures appeared to have increased concern about the availability of information. Departments were particularly protective of market-related information. There was also a natural sensitivity that by increasing the accessibility of their information they would in turn open up their decision-making procedures to greater scrutiny. This defensive approach is hardly surprising, given that the allocation of budgets in local authorities, like most organizations, is a competitive process. As a result the seemingly innocent activity of data sharing via GIS is associated by some with the erosion of control and independence.

Much of the discussion concerning the merits of data sharing focuses on the assumed inefficiencies of having duplicate data sets within a single organization (see, for example, Bromley and Coulson 1989; Onsrud and Rushton 1995). However, such arguments fail to take account of the context in which GIS development is taking place. There was evidence in several case studies that the existence of multiple data sets reflected long-standing tensions and antagonisms between departments as well as between sections within a single department. In most cases it is unlikely that the development of a GIS will resolve old scores of this type. At a more mundane level, the basic issue of convenience accounted for the wish of several departments to maintain control of their own data sets.

Economic considerations were a further matter affecting the ability of authorities to develop corporate systems. Staff in user departments were adamant that they would not contribute to systems in which they were a net inputter of data. Such circumstances were regarded as imposing unacceptable costs on departments in terms of the loss of staff time, while individual staff members resented others benefiting from their efforts. As a result it is important for the effective implementation of a GIS that the benefits to all concerned are transparent and evenly shared. This is not an issue that simply arises in relations between departments but is also evident within departments and perhaps most particularly at an individual scale.

The findings question the extent to which there is a latent demand for data sharing within the organizations studied. The results also highlight the context-less nature of the underlying assumptions concerning data sharing and raise important questions about the nature of information. The need to take account of these issues in developing a GIS are exemplified by the two case studies that had successfully achieved some measure of joint working between departments. In the first case, rather than making the data sets available, the system held lists of names of the individuals responsible for the information, with the provision of this information achieved through the exchange of disks. In the second case no data were made freely available without the written consent of the senior officer responsible for the information. Both these approaches ensure that the data set is only made available on the terms of those holding information and therefore no loss of control is implied.

These results indicate that the users interviewed during the fieldwork did not necessarily place a high priority on the ability to share information through computerized means. As a result, any attempt to develop a system involving more than one organizational unit must ensure that the data sets are carefully selected and the process of implementation thoughtfully managed.

Information requirements

This section explores the needs of users with respect to geographic information with the comments concerning data sharing providing a useful background to this discussion. Although the subject of the spatial data sets of greatest value to users varies between contexts, the findings of the case studies raise some important issues about the characteristics of the information required by users.

An interesting insight into this issue was provided by a couple of respondents in user departments who noted the wide variations in the perceptions of information between computer specialists and users. It was suggested that the technical experts regarded information as merely units of data rather than in terms of its capacity to assist with a particular task. Consequently they had little grasp of the context in which the information was being applied and therefore found it difficult to understand the needs of users. This issue was of particular significance in circumstances where the GIS project was led by the central computing section. These circumstances had a tendency to result in technical success while widespread utilization proved more problematic.

The essence of a GIS is that users will find information based on some form of geographical unit of value in their work. However, the traditional basis of most systems in organizations is financial. Consequently, with the exception of the land use planners, most staff interviewed in the local authorities studied were unaccustomed in thinking in spatial terms. These attitudes may be difficult to change as finance remains the critical determining factor in many circumstances, while the educational task involved in getting staff and the elected members to think spatially appeared likely to be considerable. Moreover, it is important not to overestimate the significance of space as in most GIS applications it is simply performing the function of an indexing system, in much the same way as presenting information in alphabetical order.

The preceding discussion shows that where a role for spatial data has been identified, very careful consideration needs to be given to the data sets that are input in the GIS. The case studies demonstrate that what may be an interesting technical exercise for computer specialists need not be perceived to be of any value by users. The constraints on staff time mean that systems will only be used that supply information essential to the work of a particular individual rather than of less certain value.

The case studies highlight a number of important issues in terms of the characteristics of the spatial data required by users. First, while staff in different departments or sections might wish to make use of the same data, the level of *accuracy* they require may differ markedly. For instance, the needs of those involved with legal documents such as land charge searches will differ greatly from those responsible for analysing changing land use patterns, despite both utilizing information on planning policies and development control. In recognition of such issues, one of the case studies had devised a system whereby a halo was placed around planning constraints such as the green belt, to ensure that non-specialists checked the precise designation with the appropriate professional staff. Such procedures acknowledge the inevitable differences in the level of spatial detail in existing data sets as well as the often critical additional information held by staff.

A similar issue to accuracy is that of *up-to-dateness*. Again the currency of the data required by users varies according to the purpose for which the information is being used. One of the most important features of the Ordnance Survey's digital maps is their up-to-dateness. However, the findings demonstrate that this was not necessarily as critical to users as generally assumed. For instance, an analysis of map usage in one of the metropolitan authorities showed that 45 per cent of those referring to the most up-to-date maps were from the planning department, another 45 per cent from the engineering and transportation department and only 10 per cent from the rest of the authority. It appears, therefore, that most staff outside the technical departments were satisfied with whatever hard copy maps were available in the department, and where this was not the case, a special one-off purchase was made.

An issue that was emphasized by a great many users concerned the *quality* of the hard copy produced by the GIS. Regardless of the capabilities of the various systems the majority of the users judged the technology in terms of the hard copy generated and therefore, by implication, the quality of the plotter. Many users were highly critical of the GIS purchased by their organization, stating that, despite the considerable expense, the quality of the maps produced was inferior to traditional reprographic methods. As a result, in some cases the acquisition of an electrostatic plotter seemed likely to have a far more significant impact on the likelihood of achieving utilization than the purchase of sophisticated GIS technology. This reinforces the notion that the image of a technology is as important, if not more important, than its technological capabilities.

Finally, a critical characteristic of geographic information is the form of *spatial referencing* adopted. The situation in one of the authorities where there

GIS and organizations

were four definitions of an address within a single department typifies the complexities of this issue. In many cases spatial referencing is viewed as a purely technical process. However, in practice the variety of approaches and definitions embrace a combination of habit, tradition, personal preference and idiosyncrasies. As a result, any attempt to modify these referencing systems will have to ensure that the benefits of change are self-evident to all. It is therefore important that decisions about this characteristic of the spatial data generated by a GIS, like those discussed earlier, must be firmly based on a realistic understanding of the needs of users.

The service provided by a GIS

A GIS is, in effect, providing a service to users. This is an important issue as it is vital that the process of supplying the needs of users does not get in the way of the provision of the information. In this regard it is crucial that information is provided when users want it and in a form that can be easily utilized. In cases where it is necessary to request specialist staff to generate the information required, the length of *time* involved in this process is vital. No information system will gain credibility if it takes so long to deliver the information requested that the decision has already had to be taken. Given that many requests for spatial data take a non-standard form, users in several case studies considered the length of time involved in developing specialist routines to be unacceptable. Furthermore, even in the case of operational activities, one of the few successful applications had found that the GIS facilities were slower than manual procedures. In this case, while the manual system did not provide the potential flexibility offered by the GIS, the priorities of management were to meet short-term goals set by the politicians. As a result there was an ongoing sense of uncertainty about the future of the GIS project. These circumstances will not always be the case, but undoubtedly the timely production of information is a crucial consideration in many organizations.

The second area of concern in terms of the service provided by GIS relates to the *form of information provision*. One of the great advantages offered by GIS is the visual display of information. The 'pretty pictures syndrome' is often derided by GIS specialists. However, comments of users suggest such attitudes to be misguided as maps can communicate information far more effectively than pages of tabulation which few people will be prepared to read and even fewer will fully understand. From the findings of the case studies it appeared that it was precisely these highly visible facilities that were most likely to gain the attention and therefore the support of elected members. The symbolic role of these images should not be underestimated. In addition to the capacity to display maps, the findings indicated that search and query facilities were seen as important. However, staff in five of the more experienced information handling authorities questioned whether there was actually a need for GIS facilities to perform this task. These individuals felt that the information they required could be provided by the existing database packages. The advantages of these systems were felt to be their far greater ease of use and lower cost.

The final issue in terms of the service that users required of GIS technology concerned *accessibility*. For GIS to be directly utilized by non-specialist staff they must be not only physically accessible but also demonstratively easy to use. The last chapter showed that the physical accessibility of GIS facilities was still relatively restricted even in well-established sites. Moreover, GIS were generally perceived as complex and difficult to use. Even the key activity of map production required a computer specialist to execute in many cases. Only two authorities had managed to enable users to undertake their own plots. In all other cases the as yet partial mapping services were provided by specialist staff, generally in the central computing department. It was also these individuals who had access to the best quality mapping facilities.

The issue of accessibility is of vital importance where the use of a GIS implies changes to existing work practices. Many staff questioned the need to change work patterns which they felt were working well enough. Inertia is a powerful and understandable force, as change implies hassle and inconvenience with few individuals motivated by a desire to break new frontiers in terms of methods of working. Experience from the case studies demonstrates that the benefits of altering the manner in which a particular activity is carried out must be self-evident to the individuals involved if their full cooperation is to be obtained. Even the seemingly minor action of changing the contact person for the provision of maps may take some time to become fully effective as new working relationships will have to be forged.

Technology

The preceding discussion has indicated that while the value of the information provided by a GIS and the advantages of the facilities offered must be clear to users, their complexity and sophistication must be virtually invisible. An important issue for the future development of GIS is that the vast majority of potential users in the organizations studied regarded such systems as difficult to use. Staff in user departments only wanted technology that they perceived to be relatively cheap to run, user-friendly and did not increase their reliance on outside agencies such as vendors or the central computing section of their local authority. Most of the systems that were being implemented by several departments failed on all these accounts.

It was interesting that staff in these relatively well-established GIS environments were beginning to question the value of multi-purpose systems. One authority in particular was in the process of reversing the idea of having a corporate GIS to the notion of having a 'plurality of GIS'. Rather than there being one central system, individual departments would purchase the most suitable software to meet their particular requirements. As a result departments such as planning and social services might purchase SPANS in order to carry out

GIS and organizations

basic spatial analysis, while the transport network would be managed using an Arc/Oracle configuration and automated mapping systems purchased to handle the needs of other departments in this respect. In an environment of cost-centre budgeting, it was evident that the service that the technology provided to user departments was far more important than concepts of data sharing or corporate decision-making.

Summary

The identification of the real needs of users and a realistic understanding of the role of geographic information in their work is a vital component to achieving the effective utilization of a GIS. The findings of the case studies cast considerable doubt on arguments that suggest that there is a latent demand for data sharing, excepting to some extent Ordnance Survey digital maps and census data. Even in these cases the form and the nature of the data required varies considerably. The size of many departments means that such issues are as much of relevance at the departmental level as organization-wide. User requirements are not limited to the type of information generated but also need to take account of the service provided and the nature of the technology. However, given the factors influencing adoption examined at the start of the last chapter, it is possible that in some cases the needs of users might take less tangible forms. In some cases the symbolic presence of a GIS may be sufficient to justify its existence. Notwithstanding this observation, an understanding of the information requirements of users is critical to both the utilization of GIS within organizations and also more generally to the development of the technology. If such issues are not considered, it is likely that GIS will become redundant and substantial resources will have been wasted. This suggests the importance of looking at the issues from the perspective of how organizations actually operate rather than a hypothetical notion of how they should.

The resource implications of GIS

An appreciation of the resource implications associated with implementing and subsequently maintaining a GIS, like other forms of technology, must be an important part of any information management strategy. It is as important, therefore, to understand the demands that the introduction of a technological innovation will place on an organization as the needs of users. The social and political aspects will be examined later in the chapter. This section concentrates specifically on resource issues. It would seem to be important that, alongside the adoption of a GIS, consideration is given to the resources necessary to secure utilization, including such issues as the existing skill base and the availability of financial support to enable the organization to draw on the expertise of external agencies. It is likely that unforeseen circumstances will call into doubt the adequacy of these provisions, but the chances of securing effective implementation will be more problematic if these issues are not considered at an early stage in discussions. The resources being devoted to the introduction of GIS technology in the case studies were analysed in detail in the last chapter; consequently it is appropriate at this point simply to highlight a number of important issues.

Like many other forms of information technology the implementation of GIS tends to be a highly resource-intensive activity. The analysis of the experiences of the case studies indicates that, regardless of the balance between in-house and external development, the financial costs, staff time and variety of skills required are considerable. These findings also suggest that the most significant costs are often associated with data capture and editing rather than the equipment itself. The existing state and format of the data to be input into the system directly affects the ease with which this task can be completed. Furthermore, for applications where up-to-date data are crucial, such costs will be recurrent. It is not surprising in the prevailing economic climate that virtually all the case studies felt that the inadequacy of staff resources devoted to GIS development was hampering progress. Even harder to deal with in many ways was uncertainty over the availability of resources in the future. It is even possible that conditions will change to such an extent as to make the implementation of a GIS untenable in the manner originally intended. For example, the introduction of direct charging for the use of the central processing unit in one of the authorities made it financially unrealistic for the planning department to continue its involvement with the corporate GIS project, as the system made substantial demands on processor time. It is, however, probable that if the GIS had been producing information that was of value to the planners, their attitude may have been different. This highlights the close relationship between user perceptions of an innovation and their willingness to expend scarce resources.

Resource issues are often advanced as a purely practical argument against the development of departmental GIS, as it is regarded as difficult for one department to supply all the skills and resources necessary for successful implementation. The three departmental systems included in the case studies provide contradictory evidence. In two instances the departments were either fully self-sufficient or had access to the necessary expertise through personal contacts outside the authority. In both cases the option of working with other departments in the organization was never considered or else positively rejected. However, in the case of the third system a lack of appreciation of the scale of the data capture exercise and an inability to cope with the technical difficulties that arose contributed to the virtual failure of the project. In this case a decision has since been taken to participate in the authority-wide implementation of GIS, in the tacit acknowledgement that this transfers responsibility for dealing with the problems encountered to others within the organization. These findings suggest that there is no reason to assume that departments, any more than whole organizations, lack the resources necessary to implement a GIS; however, it is important to identify those resources that will have a fundamental influence on the eventual outcome of the project.

GIS and organizations

The findings presented in the last chapter indicate that a wide variety of skills is required to implement a GIS. Crucial among these is management expertise, which is examined in more detail in the next section. On the technical side, most GIS need a considerable level of customization to become operational and therefore skilled systems analysts and programmers are a vital resource. In this respect the case study findings suggest that there is a much greater chance of utilization if the computer specialists are based in user departments, or are at least responsible to individuals within these departments, rather than confined to the central computing section.

Training is one of the most important ways of supplementing the existing resource base of an organization. It may take a variety of forms, including specialist courses outside or inside the organization or more incremental approaches involving 'on-the-job' guidance. Most of the specialist courses in which participants from the case studies were involved were provided by vendors. These courses tended to be targeted at computer specialists rather than users, with many of the participants expressing dissatisfaction. It was felt that they were expensive and often poorly structured and inappropriately focused. As the needs of each individual tend to be slightly different, ensuring that the level is appropriate to all can be problematic. Respondents noted adverse experiences of both courses that had assumed too much knowledge and also those that had assumed too little. In the case of users it is important that consideration is given to their individual training needs in terms of the underlying techniques and the information to be generated as well as the technology. It is inevitable that failure to provide users with the skills necessary to make use of GIS facilities will result in such systems being under-utilized. Experience from the case studies indicated that the most effective training was conducted 'on the job', as it dealt with the day-to-day activities of those involved. The personality of the individual providing the training was perhaps the most crucial aspect of the process, as perceptions of their demeanour appeared to influence utilization. Technology, like many other aspects of organizational life, tends to be personalized; as a result it was crucial that the individual providing training was regarded as friendly and supportive. It was often a fellow user who performed this task most effectively, with such individuals encouraging users to consult them when they hit problems rather than seeking some other method of satisfying their requirements. While such an approach was time-consuming in the short term, the chances of securing longer-term utilization appeared to be relatively high.

The effective utilization of GIS is not just a question of enabling prospective users to manipulate the technology. It is also, and perhaps more importantly, a matter of providing them with the necessary skills to handle geographic information. A number of respondents in the more advanced environments were concerned that while users had access to automated mapping facilities, for example, they did not at the same time possess the cartographic expertise required to make proper use of these capabilities. Similar issues apply to the application of spatial analysis techniques. Consequently, increasing accessibility to geographic information without simultaneously ensuring that the staff concerned have the necessary information handling skills could result in a deterioration in the service provided by the organization.

This discussion has highlighted a number of important issues concerning the resource implications of developing a GIS. However, it was evident in the case studies that there were a number of significant but less tangible resources that had a substantial impact on the successful implementation of GIS. These were, first, experience, and second, the ability of the organization and, by implication, the staff within that organization to cope with change. Experience of handling spatial data and developing computer-based systems was certainly advantageous but perhaps the most important element of this experience was the ability to learn from previous mistakes. Furthermore, some organizational cultures demonstrated an ability to facilitate innovation while others had a tendency to strangle such proposals. These issues will be considered throughout the rest of the chapter. However, the findings from the case studies suggest that it is undoubtedly important that the development of any information management strategy takes account of these less tangible resources. This is not to say that environments that lack these qualities should not introduce GIS, but that account must be taken of these circumstances in the overall strategy.

Summary

The findings from the case studies indicate that few authorities or departments had identified the key data sets that underpinned effective service delivery in their organization and related such requirements to the available resources. Emphasis tended to be placed on the technology divorced from the value of the information to be generated or service that such facilities would provide. The analysis of user needs in the case studies highlights the inappropriateness of many of the managerial rationalist assumptions that have been made about geographic information. In particular there was little support for suggestions that there is a considerable latent demand for sharing spatial data sets within organizations. The only exception to this is probably Ordnance Survey maps. This leads to something of a paradox in that the approach to implementation in many of the case study organizations was guided by an acceptance of managerial rationalist assumptions, yet the failure of these assumptions to be borne out in practice was rendering effective implementation problematic.

These findings indicate the urgent need for much better theoretical and practical understanding of the nature and role of information in the activities of organizations. For instance, what information is really essential to the work of organizations and what is merely nice to have? The methods selected to handle the provision of this information must take account of the available resources in terms of financial costs, staff time, skills, experience and the culture of the organization. The result of ignoring these issues are starkly demonstrated by the case studies. In at least three of the authorities with partially or fully operational systems, staff stated that they would not miss the GIS if it was removed. In one case £1 million had been spent on the development of a GIS with virtually no return.

Commitment and participation

There seems little disagreement that the chances of effectively implementing a technological innovation such as a GIS are strongly influenced by the level of commitment that exists within the organization. However, there are significant differences in the assumptions concerning the ease with which such commitment can be generated. The managerial rationalist perspective suggests that, given a logical explanation of the value of the technology to the organization, commitment to the project will automatically follow. In the case of GIS this implies the involvement of individuals throughout the organization as the main benefits of the technology are perceived to be associated with a corporate style of implementation. In contrast, the social interactionist perspective envisages the introduction of new forms of technology as likely to challenge the existing social and political order within an organization and therefore achieving widespread commitment to be far more problematic. Moreover, it will be necessary to sustain this commitment during the full lifetime of the project. It follows on from these divergent analyses that the types of mechanism that are envisaged as facilitating the development of commitment should differ. The managerial rationalist perspective places an emphasis on working parties, pilot projects and champions as the means of encouraging effective corporate working, while the social interactionist perspective suggests the need for user-centred participatory approaches. Underlying such prescriptions are questions about who should be involved in making decisions about the implementation of a particular technology.

GIS are often viewed as a multi-purpose technology which can satisfy the needs of a variety of users. As a result, commitment and participation must be achieved both vertically in terms of a variety of skills such as managers, users and computer specialists and also horizontally with respect to what in most organizations is a heterogeneous set of potential users. The following analysis therefore explores, first, the extent to which horizontal and vertical integration had taken place between the different units in the organizations studied, and second, whether there were any mechanisms that appeared to foster commitment and participation. The case study findings are examined to see if there appeared to be any tradition of joint working in the authorities and the extent to which there was widespread participation in the implementation of the various GIS. The second part explores the impact of a number of mechanisms that the managerial rationalist perspective suggests are likely to increase the chances of achieving commitment and therefore successful implementation. These include the presence of a champion, working groups and pilot projects.

The organizational context

Horizontal integration

Organizations within the local government sector in Britain are often viewed as exhibiting the worst excesses of a bureaucracy, despite such an organizational form being by no means the exclusive preserve of local authorities (see Chapter 4). The tendency towards departmentalism, for example, is a common feature of many private as well as public sector organizations. It is, however, important to examine the extent to which this characteristic could be seen in the case study organizations and its implications for the implementation of GIS. Consequently the analysis addresses whether the existing organizational form is largely an irrelevance, as the managerial rationalist perspective suggests, or whether it has a significant influence on the process of implementation.

The case study findings tend to confirm the notion of local authorities as highly departmental in nature, with only one exhibiting any indication of corporate working in the general way in which service delivery was organized. The latter was based on a strong lead taken by elected members. Respondents even in the smaller authorities stated that there was a tendency for 'family squabbles' to develop between heads of departments. The appointment of a new chief executive in three cases had resulted in an attempt to introduce a greater level of corporate working, which had been further encouraged in two instances by the consolidation of formerly separate offices into a new central headquarters. Even in these three authorities this management style had proved difficult to sustain, particularly as legislative changes were tending to promote decentralization and competition between departments rather than cooperation and cohesion. It is important that this legislation is not viewed as the cause of fragmentation as it appeared to be merely providing a justification for a pattern of working that was inherent in the vast majority of the organizations studied.

The absence of a culture of collaboration in most of the case studies appeared to be built upon and reinforced by the professional allegiances that existed within the organizations. As a result there was a strong sense that individuals valued their professional identity far more than their links to a particular authority. A professional such as a surveyor tended to perceive that they had more in common with fellow surveyors outside their particular organization than with other professionals within. In terms of education, experience and even style of language this was very often the case. These values and traditions were institutionalized within the organizations by the structure of departments. In four of the case studies the professional rivalries that inevitably result were accentuated by the physical separation of departments in different locations within the authority. These four were hosts to six case studies with all the departmental systems being introduced in these environments.

The process of allocating the annual budget frequently provides a focus for battles between organizational units, couched as it often is in an atmosphere of winners and losers. Reductions in available finance appeared in many cases to have accentuated the air of competitiveness. Personal relationships between senior staff also had a significant influence on the outcome of this process. Given these circumstances, it was possible in most of the authorities to identify an ongoing tension between the activities of the chief executive and the often conflicting priorities of individual departments.

It is not just at the organizational level that consensus can be difficult to secure. It was evident from the interviews conducted in the case studies that rivalries existed between the individual sections into which the departments were divided. Individuals within departments rarely had a full appreciation of the work-load of and pressures on staff beyond their immediate section. The by no means atypical example of the presence of four gazetteers in one planning department, highlights the largely independent patterns of working that can develop. Divisions and fragmentation appeared most likely to occur in the larger departments, particularly where a policy of decentralized local area offices had been introduced.

These findings indicate that the organizational cultures into which the GIS were being implemented exhibited little experience of corporate or even department-wide working. It has been argued by those favouring a managerial rationalist stance that, given this context, the facilities offered by GIS could provide a means of redressing these trends, leading to greater coordination and cooperation between separate organizational units. However, such comments underestimate the scale of the task involved in attempting to change the practices and values of an organization. Moreover, these findings question the extent to which the introduction of a new technology can on its own effect such a change. Discussions with individual members of staff suggest their concerns to be essentially personal, with their attitude to the technology guided by their response to the question of what's in it for them or perhaps their department. For many individuals the advantages of GIS were not self-evident, particularly where involvement in the project implied some loss of existing control. Independence was the key feature of organizational life that all departments and individuals were keen to maintain. It therefore follows that the benefits of GIS must be obvious to potential users if they are going to be willing to forego some degree of independence.

The survey evidence of the increasing adoption of GIS by departments in the context of existing corporate developments as well as the case study examples suggest a reluctance on the part of operational units to compromise their existing independence. In one case study the key individual involved was willing to incur the condemnation of the chief executive's department in order to avoid what was termed the 'corporate porridge'. Moreover, in several cases the experiences of those participating in projects involving several departments had led them to question their need for GIS facilities or to decide that they would be better off developing a separate system. The latter group tended to include departments that had a clearly identified need for GIS facilities and had sufficient skills to feel confident that they could cope on their own. Evidence from the successful case studies suggests that to maintain commitment in a GIS project involving several

GIS implementation: a problematic process

departments, the benefits of and need for the system must be sufficiently great and self-evident that any hassles and frustrations that are encountered, such as the need for compromise over a spatial referencing system, will not threaten the integrity of the project. It was also evident in these cases that each department had made a conscious decision to make use of the central GIS facilities rather than being required so to do. Given this decision, they had then purchased the necessary equipment and started to develop their own particular applications. The need for a GIS was not imposed therefore by either the central computing department or the chief executive's department. Even in these circumstances it is possible that those with sufficient skills and experience will in time decide to act independently. There was a growing feeling among users in several of the case studies, particularly those that had seen little return on their investment, that the concept of a corporate information system in a multi-functional organization such as a local authority was unrealistic.

These findings demonstrate again the importance of ensuring that the projects generated by the GIS meet the needs of users. It is the users themselves that have the fullest appreciation of their requirements and therefore it is vital that they take the central role in both the preparation of the initial information management strategy and system implementation if their commitment is to be secured. The involvement of users in the initial decision to purchase the technology also appeared to have an important influence on their subsequent attitude towards the system. Users in several of the case studies commented that they had never wanted a GIS. In these cases it was felt that the technology had been imposed from above to provide information that was of only marginal value to their own work or that of their department. It was evident that the lack of user involvement in the initial decision had caused resentment which was subsequently difficult to overcome.

Departments in many ways represent a microcosm of the trends identified in the case study organizations as a whole. None of the GIS applications being developed as part of an authority-wide project or solely by one department were fully integrated within the contexts that they were expected to benefit. Knowledge about GIS was generally confined to a tightly knit group, with information about the system often conveyed in a haphazard fashion through word of mouth. In several of the more successful cases this seeming reticence reflected a purely pragmatic realization that the staff involved could not cope with any additional demands and that to raise expectations without being able to deliver would be misguided. Many of these individuals felt frustrated by what they regarded as a lack of commitment from senior management, often accompanied by uncertainty over the level of resources that would be available. Consequently, it was emphasized that the chances of achieving and sustaining effective utilization are dependent upon the existence of a strong power base of support within individual departments as much as across an entire organization. However, achieving even departmental consensus can be problematic.

The experiences of one of the case studies indicate the importance of being aware of the motivations underlying the commitment of a department or even a section within a department to become involved in a GIS project. In this instance the approach of a vendor to make the organization a test site for its product was said to have flattered senior staff and thereby stimulated their interest. However, once difficulties were encountered, such considerations could not sustain commitment to the technology.

Overall, these findings highlight the tensions that exist at both organization-wide and departmental scales. An inherent need to maintain independence appears to have had a significant influence on the behaviour of both individuals and departments. However, one group of forces which seemed to run counter to these trends, at least for a short period, was the creation of momentum. In a few cases the GIS project appeared to have developed a self-generating momentum, resulting in a moral obligation to continue. The abandonment of such a project is often regarded as a source of organizational and personal embarrassment, calling into question the validity of professional advice. Such considerations led to the continuation of a number of the projects that were experiencing difficulties, although the resources in some cases had been reduced. In contrast, in those case studies where users were utilizing the GIS facilities, a momentum was starting to develop, leading more user departments to express an interest.

The evidence from the case studies indicates that there was little tradition of organizational collaboration, nor did it seem likely that the introduction of a GIS could in itself alter these values. Unless the benefits of the GIS were self-evident to users, it was extremely difficult to gain sufficient commitment to implement effectively an authority-wide system. The findings also suggest that such issues are replicated on a departmental scale. Commitment to a project was not something individuals appeared to give simply on the basis of the logical potential of a particular innovation. Motivations centred much more on an analysis of the personal costs and benefits of involvement. Such personal evaluations of technological innovations are less likely to be favourable where users feel excluded from the decision-making processes. Widespread participation in the implementation of the various GIS appeared at best to be limited. It is therefore important to examine the extent to which these relationships were replicated in terms of the interaction between different skill groups.

Vertical integration

The findings of the case studies suggest that for the effective implementation of a GIS, collaboration between users, senior management, elected members, technicians and computer specialists was essential. Without the commitment of these various groups, effective utilization was unlikely to occur, regardless of whether an authority-wide or departmental system was being implemented. The case studies indicate that one of the most difficult relationships is that between users and computer specialists. In all instances there were examples of considerable tension between these two groups. Users often felt excluded from the decision-making processes which were influencing the system they would be expected to use. This lack of real participation tended to reduce user commitment to the project, thereby making it even harder to achieve utilization.

In common with the findings of previous studies focusing on the introduction of information technology, it was evident in the case studies that many of the tensions between users and the computer specialists resulted from widely varying motivations and underlying philosophies (Campbell 1990a; Downs 1967; Eason 1988; Keen 1981; Markus and Bjørn-Andersen 1987; Mumford and Pettigrew 1975). One of the most fundamental issues was that many computer specialists had little understanding of the nature of the information required by users. In some circumstances there had been a complete failure to communicate with the individuals involved unable to speak the same language. This lack of mutual understanding inevitably provoked suspicion on the part of users and consequently inhibited system development. There were also basic differences in underlying philosophy concerning the organization of computer technology. Staff in the majority of user departments were critical of the centralizing instincts of the organization's computing department, while the specialists in information technology emphasized the importance of ensuring compatibility between systems.

At the heart of these differences was the sensitive issue of control. There was a strong feeling among users that they wished to utilize the technology that they perceived best matched their needs. The development of departmental systems and the formation of coalitions of user departments with significant computing experience working together emphasizes the conscious efforts being made to avoid what was regarded as unnecessary compromise. In many respects, technological and legislative developments were supporting the demands of users for greater independence. Progress in personal computing had undoubtedly presented users with an opportunity that many were keen to exploit, while the introduction of compulsory competitive tendering for computing services enabled user departments to seek technical support from outside agencies. One of the case studies had already done so for GIS support. Despite this context, many users felt that there was a reticence, particularly on the part of senior staff in computing departments, to embrace these developments. Consequently tensions persisted.

In some cases GIS technology appeared to accentuate the difference in philosophies between users and computer specialists as it offered a new way of organizing data based on geography. For the more junior staff this was often seen positively as a potentially exciting area in which to work, while for many senior staff GIS represented a challenge to existing practices.

Like many forms of computer technology the implementation of GIS in the case studies demonstrates that there is often a tense and potentially destructive relationship between users and the computing specialists. Which set of interests should be paramount in any dialogue between these groups is therefore critical. Evidence from the case studies strongly suggests that users should be the final arbiters in the implementation process, for if the system fails to meet their needs the technology will become redundant. In instances where responsibility for GIS development rested with the computing specialists there seemed to be a high probability that, despite technical success, the application would fail to be used. All the systems that were technically operational but not yet in use were led by computing specialists. In these circumstances potential users often commented that the applications that had been developed were only of marginal value to their work and complained that there had been little consultation. This sense of exclusion emanated from their lack of involvement in the initial decision to purchase the GIS. This in turn led to a vicious circle whereby users had no commitment to the project and therefore allocated few resources, particularly in terms of staff time. This left the computing section to assume what the requirements of users might be. These assumptions often proved to be inaccurate. In such cases, while the computing staff were essentially well intentioned, they underestimated the importance of direct user involvement in system development. In contrast, where the GIS projects were led by user departments, there was a far greater chance of achieving effective utilization. However, there was one important exception to this general pattern. This concerned an authority where the lead had been taken by the computing department but the staff in that department were regarded as sympathetic to the needs of users and also aware of the misgivings that many individuals had about computers. In this environment a largely constructive relationship had been developed between users and the computer specialists which had made a significant contribution to the effective implementation of that system.

The findings emphasize that the relationship between users and computer specialists has a fundamental influence on the likely outcome of GIS implementation, both in terms of its impact on the broad strategy adopted as well as the more detailed matters of system design. Working relationships between staff in the other skill groups were generally less problematic but no less vital. The commitment of the senior management and elected members to GIS implementation was crucial as they were responsible for allocating resources. The level of involvement of elected members depended on the context. In some of the case studies the role of elected members was limited to sanctioning expenditure in committee, while in a very few cases they took an ongoing interest in system development. Greater involvement of elected members appeared to make success more likely although a lack of direct involvement by elected members by no means precluded effective utilization. It seemed to follow that the more directly members were involved with the project, the more likely they were to grant the additional resources that were frequently needed to sustain the implementation process. In many instances the attitudes of elected members were influenced by their relationships with senior staff, particularly the level of respect and trust that existed between key politicians and their senior managers. Given the relatively long time span over which investment and therefore commitment in GIS must be sustained without any great return, such relationships were often extremely important.

It will be clear from the preceding discussion that personal relationships and even rivalries between senior staff can have an important bearing on the level of commitment that individuals and departments express in GIS. In an effort to

GIS implementation: a problematic process

avoid potential problems at this level it has been suggested that it is best to work through users in middle management. It was evident from the case studies that while this may be satisfactory for detailed decisions, staff at this level tend to lack the necessary scope of vision and influence to secure effective implementation.

Summary

The findings of the case studies suggest that achieving commitment among all the various areas of expertise involved in GIS implementation can be problematic. This is particularly so where the benefits of system development are not immediately self-evident. Commitment to the various GIS projects seemed to increase where users were not only given scope to participate but had responsibility for determining the most appropriate course of action. The advice of computer specialists is undoubtedly important but in the end it is the users who are most aware of their needs. The final part of this section explores the value of a number of mechanisms that have been proposed as facilitating the process of developing commitment and participation in GIS implementation. These include the role of champions, the contribution of working groups and the use of pilot projects. The GIS literature has tended to present these mechanisms as part of the managerial rationalist cookbook but it is also possible to conceptualize them in a more participatory mode as part of a social interactionist approach. The following section examines their worth in terms of the experiences of the case study organizations.

Mechanisms facilitating commitment and participation

Champions

The concept of a champion as the key determinant in achieving successful implementation of a GIS has gained increasing prominence in recent years (see Chapter 3). The term 'champion' is usually applied to an individual who is responsible for gaining approval for the purchase of GIS technology and then coordinates the implementation of the system. These individuals are seen as charismatic personalities who can transform a seemingly apathetic environment into an innovative and progressive organization. The language used to refer to such individuals implies success born of an ability to take on existing prejudices and secure resources for their particular area of interest. The role of champions in terms of GIS implementation is usually viewed in a mechanistic sense as the solution to organizational inertia and the safeguard against failure.

The findings of the case studies cast considerable doubt on the value of the concept of champions as generally presented. The results suggest that a collection of qualities are required among the personnel involved in system development which is virtually impossible to encapsulate within a single individual. The case studies indicated that there was a need for an individual who could fight the inevitable political battles in order to ensure that sufficient resources were allocated to the project. However, these qualities were not enough on their own to secure utilization. Moreover, where the individual concerned was not a senior member of staff they tended to lack the necessary power base to succeed even in these terms. The experiences of the case studies demonstrate that it is impossible for members of middle management to win political battles both within and between departments without forging strategic alliances with influential members of senior management. It is also important to recognize that the implementation of GIS, in common with other computer-based systems, is not simply a high-profile activity. The integration of the technology into the daily work practices of an organization is dependent upon the devotion of a considerable amount of time and skill on far less glamorous activities including the specification of user requirements, data collection, technological frailties, data structures and training.

Analysis of the case study experiences emphasizes that the more successful environments were noted for having a collection of highly able individuals who possessed the full range of skills mentioned above, rather than being distinguished by the presence of a champion. It was possible in three cases to identify an individual who appeared to have made a disproportionate contribution to the implementation of a GIS in that context. Out of these three, two had achieved utilization of the GIS they were introducing. However, it was evident that such an outcome would not have been possible without the very considerable support of several other members of staff. It did not seem that the crucial feature of these contexts was the presence of what may be described as a champion; of much more profound significance was the underlying organizational culture which encouraged individuals to make the most of their skills and talents. Overall the existence of a champion did not guarantee success nor their absence failure.

The findings of the case studies not only indicate the role of a champion to be limited but that suggestions that propose such individuals as the solution to the problems of an organization may well be dangerous. Champions are likely to be good at instigating interest in a project but, on the other hand, relatively poor at dealing with the less glamorous but equally crucial aspects of system implementation. Furthermore there seems to be a tendency for these individuals to move onto other projects or even leave once the initial high-profile tasks have been completed. If this individual leaves there will be a loss of expertise as well as the key protagonist for the system. The departure of any member of staff who is involved with the implementation of a system is likely to have adverse consequences but if, as the concept of a champion generally suggests, the whole system revolves around this individual, it is probable that the future of the whole project may be put at risk.

The implementation of a GIS is dependent upon effectively combining a variety of skills and personalities. This inevitably means making the best of the strengths while accommodating the weaknesses of the individuals involved. The findings suggest the whole concept of a champion to be flawed and in many ways misleading. Effective implementation does not appear simply to depend on the presence of a champion. Moreover, there is an inherent risk in any strategy based on a single individual. Champions may make a contribution to securing commitment to a GIS project but much will depend on the personality of the individual and their willingness to listen and take account of the existing organizational culture.

Working groups

Working groups or working parties are frequently cited as ways of achieving widespread participation in projects that involve multiple interests. With the exception of the departmental systems, two-thirds of the remaining case studies had established working groups on GIS. The function of such groups was seen as coordinating implementation and raising awareness among users. However, there was a widespread feeling that such groupings tended to degenerate into 'talking shops' which slowed down rather than facilitated development and for some this justified not forming working parties. Assessments as to the value of such groupings depended entirely on the extent to which individuals believed that they could influence the course of events. In the majority of cases the meetings were viewed as largely cosmetic, failing to provide a real opportunity for the varied interests to participate. It was the users that felt more alienated, particularly where the discussion centred on technical matters which meant little to them, yet the decisions being made directly affected their work.

Working groups undoubtedly have a role in facilitating participation and thereby commitment in the introduction of a GIS. However, the formation of a working group on its own will not engender commitment. Careful consideration needs to be given to the goals and manner in which the group operates. Meetings that are viewed by those invited to participate as little more than token gestures are likely to provoke more negative attitudes than if no meeting had taken place at all. The experiences of individuals in the case studies suggest that working parties that empower users so that they feel able to influence the decisions being made have a significant impact on the chances of the technology being utilized. Unfortunately such an approach appeared to be uncommon in the case studies, despite these issues being as important in a departmental context as organization-wide. Overall concern about the matters raised at such a forum slowing down the implementation process are less crucial than that the resulting systems meet the needs of users. Delays will be inevitable if a system is developed, no matter how speedily, that no one wants.

Pilot projects

Pilot projects provide a means of demonstrating the potential benefits of, for instance, a GIS to users as well as enabling computer specialists to learn about the technology. For example, a pilot might be undertaken for a small geographical area as a precursor to the development of a system covering the whole area administered by a particular local authority. They therefore serve both a technical and organizational function. In organizational terms the intention is that the visible success of a pilot will foster the commitment of users and encourage others to investigate the potential of the system, while for the computer specialists the aim is to enhance their knowledge of the technology. However, in practice very few of the case studies undertook pilots. Several respondents could not see any value in such activities in either technical or organizational terms. They stressed that the applications being developed would still not be operational while many of the important technical considerations such as response times could not be properly evaluated with only a partial data set. As a result there would be a chance of raising expectations without learning about the limitations of the technology. In two cases, including one of the more successful implementations, it was emphasized that elected members would not support pilot projects as they regarded such activities as 'playing'. In their view either an application was needed or not. Consequently, despite the apparent value of undertaking pilot projects only one of the case studies had employed the technique. This application had not been completed but success looked likely. The reasons for this, however, could not be attributed directly to the pilot project.

There are both symbolic and rational grounds for envisaging merit in undertaking a pilot project. However, in practice the case study organizations demonstrated a reluctance to embrace such an approach. Many of the organizations studied conceptualized themselves to be following an incremental implementation strategy. This tended to involve an application-by-application approach but rarely were pilot projects seen as integral to this process. One important issue raised by the findings was that, despite acceptance of the possible benefits of pilot projects, respondents recognized that they were not part of the way that things were conducted in their particular organization. As a result it was felt such initiatives would prove counter-productive.

Summary

The findings of the case studies demonstrate the need for the commitment and participation of staff throughout the organization in the implementation of GIS. The existing organizational context in the local authorities studied appeared to inhibit rather than facilitate joint working between departments, while tensions existed across skill areas, particularly between users and technical specialists. There seemed to be little tradition of collaborative working which parallels the limited demand for data sharing noted earlier. Moreover, it did not appear that the introduction of an innovation such as a GIS would, on its own, alter the organizational environment, rather, the technology was more likely to be shaped by the organizational culture. Evidence from the case studies also points to the importance of users taking the decisive role in system implementation. Users are the group most directly affected by the development of computer-based systems yet, as the case studies show, they are all too frequently excluded from the decision-making processes that determine the nature and organization of the resources they are expected to utilize. These issues are not confined to the organization-wide systems. The departmental systems also demonstrate the importance of widespread commitment and participation to the conduct of the implementation process. Consequently, discussions focusing on the merits of top-down as against bottom-up strategies appear misguided as successful utilization requires a combination of strategic and operational thinking. Without senior management support most projects are doomed, while a lack of understanding of user interests will lead to wasted resources. This points to the importance of securing all-round commitment to the project. The findings suggest that, contrary to the managerial rationalist assumptions, commitment is not simply created by logical argument. Support for projects such as the implementation of a GIS is determined by individual assessments of the personal implications of such a development and the extent to which it can be embedded within the values and practices of a particular organizational culture. This confirms the social interactionist position that the implementation of any computer-based system must start from an evaluation of *how* an organization actually operates and not how it *ought* to, given ideal conditions.

The experiences of the case studies also emphasize that there are no quick fixes to achieving commitment and participation. The research findings question the significance attached to the whole concept of champions. The notion of an individual who possesses the full range of skills necessary to implement a GIS appears to be flawed, while perpetuating such a solution is potentially dangerous as reliance on any one individual places the system in a very vulnerable position. Moreover, a champion will only be able to thrive and achieve effective utilization if they are in tune with the prevailing organizational culture. Attempts to 'shake up' an organization usually result in frustration and disillusionment on both sides. The effectiveness of mechanisms such as working groups and pilot projects appeared to depend very much on the manner in which they were applied. In the case of pilot projects, for example, such approaches may not be acceptable to the wider organization, while the benefits of working parties depend on the extent to which individuals feel that they have a real involvement in the implementation process. These findings suggest that the priorities of users must be the focus of the implementation process if utilization is to be achieved.

Inherent within the analysis of the mechanisms that have been cited as contributing to achieving widespread commitment and participation in organizations has been the identification of a more enduring quality among a few of the case studies, namely an innate ability to innovate. Present within a minority of the organizations studied, this feature is crucial, as it appears to influence the level of commitment to the project and therefore the chances of effective implementation. If such organizational cultures exist, they must be able to sustain the development of an innovation despite changing external and internal circumstances. The next section starts to explore this concept in more detail, focusing in particular on the impact of change and instability on the development of GIS technologies.

Instability

Instability is an inherent part of any organizational context and with instability comes the often unsettling effects of uncertainty. The very essence of management is the manipulation of these forces, whether that is in coping with instability in the external or internal contexts or in actually attempting to induce a change in organizational practices. Regardless of the source of the instability, each individual will assess the extent to which the implied change threatens their position and activities or offers them new opportunities. No change is neutral. It is always accompanied by both benefits and costs and therefore personal threats and opportunities. This mass of individual judgements are at once influenced by the organizational culture and also simultaneously contribute to the development of that culture and hence the overall attitude to technological innovations such as GIS.

The implications of change and instability are likely to be considerable for the implementation of GIS technologies. The introduction of GIS necessarily implies some change in the organizational traditions of handling spatial data. The extent of this change will depend on the type of application being developed and the values and experience of the organization. At the same time any project is subject to the vagaries of an ever changing environment and therefore modifications to priorities. The often considerable lead time required to secure an operational application places GIS technologies at risk of being superseded by what are seen as more urgent matters. Furthermore, sustaining a project which will undoubtedly face problems and difficulties during the course of implementation may prove problematic. This raises the important question of how organizations cope with instability in general and most particularly with respect to technological innovations such as GIS.

For the managerial rationalist approach, dealing with instability is largely a matter of pre-planning as such circumstances are envisaged as readily controlled. However, the social interactionist perspective views change and instability as less predictable with each organization responding differently. The following discussion explores the merits of these perspectives. It briefly outlines the pressures for change being encountered by the case study organizations, examines the impact that this instability was having on the process of implementation and considers how the environments studied were coping.

Pressures for change

Pressures for change can take one of two forms. The first concerns what may be termed standard or background instability. This refers to the ever present aspects of organizational life such as minor legislative changes, internal alterations to existing organizational structures, or the resignation of staff, all of which will be expected to take place but when and in what form is uncertain. The second are less predictable but no less regular events such as major governmental reforms or an economic crisis. Both these types of instability may have short- or long-term implications, although the latter tend to be regarded as the more disruptive. Overall, much appears to depend on timing and the existing circumstances within a particular organizational context. The sources of instability may be external or internal to the organization; however, in both instances the crucial issue is how such pressures are internalized by staff within the organization. The major sources of instability experienced by the case study organizations came from central government and the impact of the recession of the late 1980s, including the associated pressures that this placed on vendors. In addition there was an ever present background of change. Each of these will be examined in turn.

Central government

Legislative changes have a significant impact on the work of all organizations. It was noted in Chapter 4 that local authorities in Great Britain have experienced regular changes to the manner in which they are organized. Modifications to the number and responsibilities of local authorities have usually been accompanied by initiatives proposing changes to the structure and arrangements for internal management. The current period of reorganization is no exception, with the review of the distribution of local authorities taking place alongside the introduction of deregulation in the form of compulsory competitive tendering. As a result, local authorities are no longer seen as service providers; rather, their role is one of enablement (see Chapter 4). It is inappropriate at this stage to dwell on the changes themselves; rather, the subsequent analysis will explore the implications of these developments for the organizations studied and most particularly the implementation of GIS.

Major legislative changes are a recurrent part of the local authority context. However, for those working in such environments each set of proposals has to be assessed for their personal and organizational significance. Such circumstances have a tendency to divert resources away from new policy or operational initiatives towards defending the viability and traditions of the organization. In terms of GIS these conditions made some politicians and senior managers wary of making an investment in a technology that might not yield any benefits for some time. Similarly, a number of those already implementing such systems found resources, particularly staff time, diverted to what were regarded as more immediate priorities. Instability in the organizational context as a whole therefore has implications for the detailed development of projects such as GIS.

The most tangible aspect of the move towards enablement in the local authority sector for implementation has been the widespread introduction of a cost-centre-based management structure. Alterations to the manner in which services are delivered, accompanied by changes to the structure of an organization, inevitably affect the role and flow of information. All but one of the authorities studied was either in the process or had already undergone a major programme of internal reorganization. In some of the Conservative-controlled authorities the GIS project had undergone a complete change of philosophy from a centrally directed development to a much more dispersed approach based on an ideological commitment to opening up service provision to the private sector. At a practical level, reorganization of this nature leads to a redistribution of tasks and responsibilities which in turn alters working relationships. For example, one of the computer specialists commented that it was impossible to know from one week to the next who were the appropriate people to consult in user departments. In some cases the transfer of a key enthusiast to other activities had slowed the development of a particular GIS application. In one instance, reorganization had resulted in the separation of an important area of application from the staff with GIS skills as an existing department was split in two.

The experiences of the case study organizations also demonstrated the extent to which change provokes further changes that would not otherwise have taken place. For example, it was evident that a significant proportion of senior staff had felt disinclined to face the political battles associated with the process of reorganization and had therefore taken early retirement. This appeared to create a sense of uncertainty and muddle, at least in the short term, particularly where the individual concerned was well respected within the organization and had supported GIS developments.

The introduction of a cost-centre-based structure had already had direct implications for the management of information within several of the case studies. Traditionally the collection of data by any one section within a local authority was regarded as being undertaken on behalf of the whole organization and therefore should be freely available to all. Many authorities were finding such an approach difficult to sustain in an environment of client–contractor relationships. Decentralized budgets were encouraging those generating particular data sets to seek a return on their investment commensurate with the cost of production or added value. Respondents in all types of authority were either already experiencing or expected intra- as well as interdepartmental charging for information to be an inevitable result of this process. Moreover, some sections were looking to outside suppliers for information. Consequently, cost-centre-based managerial structures generally reinforced existing tendencies towards fragmented patterns of working.

The findings suggest that the climate of change created by major legislative changes had implications for the implementation of GIS technologies in the case studies. However, the response of no two organizations was exactly the same, with the majority party not having a significant influence on the overall approach. This raises issues about how organizations cope with change and uncertainty which will be returned to after examining the other main sources of instability.

The recession

Prevailing economic conditions colour perceptions as to the most appropriate course of action to follow. In addition to the climate of financial restraint created by the recession of the late 1980s and its fall-out into the 1990s, it had two significant implications for the development of GIS technologies in the case studies. The first of these concerned the associated collapse of the property market which led to a reduction in local-authority-sponsored building as well as less private sector construction needing to be administered and checked. In such circumstances staff failed to be replaced and new initiatives were cancelled in the very departments most likely to be involved in GIS development. A second, more indirect impact of the recession was the difficult market conditions created for companies in the electronics industry. As a result, virtually all the case studies encountered changes in the vendors from which they purchased their original software. In some cases the vendor withdrew support from the product, in others companies merged, while in general there appeared to be a reduction in development programmes and the extent and availability of support staff. Given the instability that such circumstances create, two respondents commented that the economic viability of vendors is in many ways more important than the quality of the product they provide.

Background instability

All organizations experience constantly changing circumstances. Most of the destabilizing influences on the activities of organizations, when taken on their own, appear little more than minor inconveniences. However, the cumulative effect of such developments can have a significant influence on the viability of a project such as the introduction of a GIS and even the survival of the organization. In terms of the experiences of the case studies with respect to GIS implementation, important sources of instability included: uncertainties over the annual budgeting process, legislative amendments, local elections, staff departures and the vagaries of external agencies, most particularly vendors. Each of these will briefly be examined in turn in order to give a flavour of the circumstances encountered by the case study organizations and the impact on implementation.

Each local authority, like most organizations, goes through an annual budgeting process which determines the allocation of resources to departments and projects. During the 1980s central government increasingly took control over the level of finance available, decreasing local discretion and restricting spending. Furthermore there was often considerable uncertainty over the level at which annual budgets could be set, thereby making planning difficult. For a project such as GIS which requires a relatively long period of investment before yielding results, such circumstances proved problematic. Managerial rationalist assumptions suggest project planning to be a relatively straightforward activity. However, in times of financial uncertainty such tasks are difficult. Several of the case studies, for example, had the resources allocated to GIS development reduced from the original levels agreed during the course of implementation.

All organizations tend to exhibit a certain ritualistic element in the way that they conduct their affairs. The budgeting process is usually one element of this but alongside the financial aspects of organizational life is usually a programme of key meetings and events. In the context of the organizations studied, local and national elections tend to have a significant influence on patterns of working as, for instance, gatherings of share-holders have in the private sector. In virtually all the case studies, impending elections resulted in a large proportion of decisions being deferred. Even in authorities where the majority party was likely to remain in power, similar attitudes were encountered as the responsibilities of the leading politicians were likely to change and therefore the emphasis of particular policy areas. New relationships would also have to be forged between senior staff and politicians. In some cases all development work on the GIS was halted during the run-up to an election.

Modifications to existing legislation also provide a constant source of disruption to both existing computer-based systems and those under development. For example, a seemingly minor alteration in the form of standard statistics required by a government department or the institutions of the European Union will result in systems having to be modified. A good example concerned the changes made by the Department of Transport to the specification of the road maintenance and management system which they require local authorities to keep. The highways department in one of the case studies had perceived GIS technology as the most appropriate method of holding the information required and therefore embarked on the development of a GIS-based application. However, part way through what was proving a problematic process, the Department of Transport altered their specification without providing sufficient additional funding to cover the costs of GIS-based systems. This in many ways proved the last straw. The GIS was abandoned in favour of a solution based on a database management package which gave the option of adding a graphical front-end at a later stage. In certain circumstances modifications to existing legislation may stimulate new developments. However, once work has been started or a system established, such changes are potentially disruptive.

The departure of personnel is an inherent part of any organization. It is potentially destabilizing, although it also offers the opportunity for the introduction of a fresh face and new ideas. In most of the case studies the team of staff most directly involved with the implementation of the GIS had been reasonably stable. However, in one case the resignation of the instigator of the project at the same time as an organization-wide restructuring programme was taking place, led to a near-critical loss of momentum. At best the departure of an important member of staff delays progress as the replacement will inevitably take time to find their way round the organization and the system. In most cases staff departures lead to a re-evaluation of priorities and in some cases this may lead to a change of direction or emphasis. Overall, there was a sense within the case studies that nobody was irreplaceable, but that some individuals may be more difficult to replace than others.

The external agencies that caused the case studies most disquiet in terms of GIS implementation were the vendors. In the majority of cases the expectations of users were not matched by the support they received. For those with limited computing experience these expectations were often unrealistically high. Nevertheless, perceptions are often more important than the detailed agreements and failure to provide what was regarded as adequate post-sales support had a destabilizing impact on a number of the projects. Those involved with the smaller GIS installations felt particularly undervalued and frustrated. Further disillusionment was caused by what was often seen as a conscious effort on the part of the vendors to create product instability and therefore ensure an ongoing and perhaps even increased share of the market. The production of regular updates to existing products reflects both technological progress and a sensible marketing strategy on the part of vendors, for new software often requires enhanced hardware. A number of the case studies resented this continuous drain on resources, particularly when support was withdrawn from previous versions of the software. In one case in particular the need to pay a six-figure sum for a new interface within the existing software proved the final straw for what had been a largely unhappy experience. More cynical tactics on the part of vendors involved poaching key staff and attempting to disrupt development of a competitor's product. The latter tended to occur in the larger authorities, with the newcomers offering favourable terms to a department not currently involved in the existing GIS project in the hope that other sections would in time switch to their product. The ultimate aim therefore was to secure the whole organization. The activities of the vendor community are not surprising, given that their overall aim is to survive in a highly competitive market. However, from the perspective of the case studies, they often represented an additional layer of instability and uncertainty. The involvement of some of the organizations studied in user groups and attempts to ensure that new contracts had clearly specified targets demonstrated the increasing efforts being made to reduce the potential for instability from this source.

Implications of instability for GIS implementation

Instability is both an inherent part of any organizational context and also has a highly significant impact on the implementation process. The preceding discussion has outlined the main sources of instability experienced by the case study organizations and provided examples of their destabilizing effects on GIS implementation. Overall, however, the striking feature of the findings was not these individual incidents but rather the extent to which each organization exhibited a unique approach to handling change. Each organization had devised their own manner of coping with instability based on the culture's traditions and values which was in turn reflected in the behaviour of individual members of staff. Consequently, the impact of instability on the implementation of GIS technologies mirrored the approach adopted to other forms of change. For example, it was evident that one of the case studies that had achieved utilization dealt with change by ensuring that its effects were kept to a minimum. This was evident both in the way that the introduction of GIS technology was handled and in their approach to the legislative changes introduced by central government. As a result, this organization stood out from the other studies for apparently being far more stable. The GIS applications that had been developed were based on information necessities, with the system itself replicating as closely as possible existing approaches to data handling. Similarly the reorganization of the internal management structure was a far less significant matter than in the other case studies. While this might not seem to result in the most exciting or interesting initiatives, the technology was being used and there seemed every prospect that the project would be sustained and developed for some time to come.

The opposite approach to change from that outlined above is where the organizational culture is such that new initiatives appear to be positively relished. Change in such environments is viewed in terms of the opportunities it presents rather than as a potential threat to the existing order. Two of the organizations studied exhibited these characteristics. One had already achieved the utilization of several applications while the other was nearing completion of its first. Individuals within these contexts seemed keen to embrace new ideas, leading to a sense that the organizations could be described as innately innovative. Again this culture was reflected in every aspect of their work with new policy initiatives being undertaken alongside managerial and technological developments. In terms of GIS implementation there was an expectation that problems would be encountered and circumstances would change, but rather than threatening the integrity of the whole project, such incidents were regarded as part of a learning process which would yield beneficial results in the end. There was therefore an innate confidence within the culture that if you give capable individuals sufficient space they will deliver.

The majority of the case study organizations appeared to be controlled by changing circumstances and events rather than exploiting them to their own advantage. There was a tendency for commitment to a new project to dissipate when problems were encountered. As a result, once the process of implementation became problematic and the rapid delivery of what were often highly ambitious plans looked unlikely, there was a tendency to take fright even where there had been a willingness to take the initial risk of purchasing a GIS. This often led to resources being reduced, thereby compounding the difficulties facing the project. The fragile nature of commitment in these contexts poses problems for initiatives such as the introduction of GIS technologies. The long lead time associated with the development of operational applications makes projects of this nature highly susceptible to the destabilizing effects of changing circumstances. The logical approach to the introduction of a GIS in the face of uncertain long-term commitment would seem to be the development of a very simple application, such as utilizing the technology as a mapping system. However, there was a tendency to view such applications with contempt, leading to ambitious plans which frequently appeared destined to come to nothing. The paradox in such cases was that, without the ambitious plans, organizational support for the initial adoption of a GIS was unlikely. A crucial aspect of the implementation process, therefore, is the ability of organizations to sustain the development of the technological innovation in the face of changing internal and external circumstances.

Summary

The findings of the case studies suggest change and instability to be both a highly significant component of the overall context and also part of the process of GIS implementation. In many ways the only thing that is certain is that circumstances will change, although when and to what extent is the conundrum that confronts those involved with the introduction of GIS technologies. There is little evidence from the case studies that the destabilizing effects of change can be readily controlled through the formulation of rational strategies. Change and its impacts appear, therefore, to be less predictable and far more difficult to manage than the managerial rationalist assumptions suggest. The response of individual organizations to change seemed in practice to reflect deeply entrenched cultural values and traditions. While each organization had its own unique approach to instability, it was possible to identify three separate types of response: those that attempt to minimize the effects of change and thereby remain essentially stable; those that are innately innovative and treat change as a catalyst for creativity; and finally those that, while initially embracing change, find it difficult to sustain this process. This suggests the important role that organizational cultures have in the process of securing the effective implementation of GIS technologies.

Conclusion

The findings of the case studies point to the complex nature of the process of effectively implementing technological innovations such as GIS. More importantly they also cast doubt on the underlying assumptions that have tended to dominate the GIS literature. The experiences of the organizations studied indicate that achieving a technically operational application by no means guarantees utilization. Moreover, the technical performance of a particular product did not appear to have determined the outcome of the process. Consequently, implementation appears to be a largely social and political process which has to be nurtured and cajoled rather than being imposed and controlled. Managerial rationalist assumptions about the latent demand for sharing spatial data within organizations and the associated necessity for corporate working have so far largely failed to be fulfilled in practice. Such attitudes and experiences are highly significant. They are not a symptom of irrational behaviour on the part of well-meaning but essentially ignorant individuals. Rather, these actions were guided by a series of often well-considered rationales which reflect the practicalities of organizational life. Overall these findings emphasize that if an innovation such as GIS is to be utilized, the starting point must be an understanding of the values and practices of the particular organization and not an idealized conception of how it ought to operate. These issues were evident in relation to one of the most important aspects of implementation: coping with change. The experiences of the case studies indicate that the underlying culture of the organization was reflected in the manner in which individual members of staff dealt with the uncertainties resulting from instability.

These findings suggest that the explanatory framework offered by the social interactionist perspective equates most closely with the experiences of the organizations in this study. The implementation of technological innovations, therefore, is not merely a technical process, nor one that simply requires the formulation of rational management strategies. Rather, it is a social and political process which takes place within unique organizational cultures subject to ever changing conditions. The theoretical and practical implications of these findings will be examined in the final chapter.

CHAPTER 8

GIS: innovation or irrelevance?

Introduction

There is an increasing recognition of the complex nature of the problems confronting societies throughout the world. Media coverage of famine, poverty and destitution at home and abroad provide stark illustrations of the inadequacies of existing responses to these fundamental human concerns. The insecurity that such images create has led to an underlying sense that there must be a way of resolving these problems more effectively than has so far been achieved. It is into this vacuum that technological innovations, especially computer-based systems, have been viewed as having a potential contribution to make, even providing the solution. Much of the folklore surrounding computers endows them with the capacity to compensate for the inadequacies of human intellect and thereby enhance economic and social well-being while at the same time ridding societies of the social deviance that threatens their integrity.

This sentiment is reflected in much of the current writing concerning geographic information technologies. Claims such as those embodied in the Chorley Report (Department of the Environment 1987) commissioned by the British government, suggesting that the contribution of GIS should be set alongside that of inventions such as the microscope, telescope and printing press, appear to provide considerable grounds for optimism. It has not been the aim of this study to investigate the moral dimensions of computing technology for society, fundamental and crucial though these issues are; rather, it has been to explore how users are responding in practice to the challenges and opportunities that such systems present. An important characteristic of GIS is the extent to which they are an organizational rather than a personal technology. GIS are not usually purchased by a single individual for their sole consumption. Frequently, their acquisition has to be negotiated among competing interests within an organization and utilization is dependent on the acceptance of its value in social and political terms. As a result, this study has sought to examine the relationship between a technological innovation, namely GIS, and organizations, in this case contexts subject to the framework provided by British local government.

The process that is responsible for transforming a collection of equipment that represents little more than untried potential into a taken-for-granted part of the everyday activities of an organization is implementation. This process has therefore been at the heart of this investigation. The very essence of implementation is change. It is impossible to envisage circumstances in which implementation, be it that of a new technology or policy, does not imply change. As a result the term 'implementation' is increasingly becoming synonymous with the management of change; a phrase that might be more accurately expressed as the cultivation or nurturing of change. The objective of the research has not been to provide users with ready answers to the conundrums that the implementation of GIS technologies confronts them with on a daily basis. There are none. Rather, the premise is that through enhanced understanding of the social and political processes that affect the relationship between organizations and technology, individuals within those contexts will be better able to influence their own destinies. An important aspect of the research therefore has been to examine the working assumptions that dominate the debate concerning the utility of GIS. In order to provide a framework for analysis, three perspectives on implementation were devised, namely technological determinism, managerial rationalism and social interactionism. Each embodies a differing range of assumptions which seek to explain the experiences of users and is accompanied by very different prescriptions as to the most appropriate approach to employ as a foundation for implementation. This has proved a valuable analytical tool and will be drawn upon in subsequent comments.

The following discussion examines the implications of the research findings, first in terms of their contribution to theoretical understanding of the relationship between technology and organizations, and second with respect to the practical aspects of GIS implementation. The remainder of the chapter emphasizes the urgent need for further investigations that explore the experiences of users in real-world situations and considers whether GIS technologies should be termed 'innovation' or 'irrelevance'.

Implications of the research findings

Theoretical developments

One of the most striking features of the research findings is the problematic nature of the process of implementation. Even for those organizations that had achieved some measure of utilization, continued use was by no means certain and the process itself had brought with it both costs and benefits. Implementation in the case study organizations was not a linear process, having a clear start and

GIS: innovation or irrelevance?

finish and progressing inevitably through a series of stages. Rather, implementation appeared to be a process of learning where not all the participants derived the same conclusions from events, learnt at the same speed or even felt sufficiently motivated to become involved. These findings are by no means new. The problematic nature of implementation and the mixed results experienced have been noted in relation to computer-based systems in general as well as other fields including policy initiatives (see, for example, Barrett and Fudge 1981; Eason 1988; Elmore 1978; Kearney 1990; Lyytinen and Hirschheim 1987; Moore 1993; Mowshowitz 1976). The important issue for this research was to explore the reasons that the case study organizations encountered difficulties in implementing GIS technologies and thereby to provide some insights into the underlying processes. In particular the research focused on enhancing understanding of the nature of technology and the extent to which assumptions about the critical features of organizations, including the role of information in decision-making, influence implementation.

The findings of the research suggest that technological innovations are perceived by users in terms of their social and political meaning rather than as a particular configuration of equipment. GIS are therefore viewed in relation to a particular application such as a property management system rather than as a computer-based technology capable of storing, manipulating and displaying spatial data. More importantly, individuals within each organizational context tended to make judgements about such systems in relation to whether they offered potentially fruitful opportunities or threatened their existing position or status. Given the social and political dimensions of the technologies, each of the organizations studied appeared to be reinventing the particular system purchased. Such a process should not be derided as highlighting the inadequacies of users. It is not an issue of organizations reinventing the wheel, but rather a process of ensuring that the wheel fits the particular cart to which it is to be attached. Reinvention is therefore indicative of the close relationship between the introduction of a new technology and organizational learning.

The evidence from the case studies suggests that technologies such as GIS are very much more than units of equipment. They are socially constructed, thereby accruing their meaning from interaction with individuals in a particular organizational context. These findings confirm earlier studies focusing on the experiences of users with respect to computer-based systems and innovations more generally (see Bijker *et al.* 1987; Goodman *et al.* 1990). Consequently, technologies appear to consist of a combination of equipment, set of underlying techniques and knowledge from which to derive an understanding of the value and role of such systems. It is implicit that in practice a technology such as GIS embodies a vast array of perceptions rather than simply being viewed as a single entity. It is immaterial to most users whether the system they have purchased equates to the most widely held definition within the academic community. It is its symbolic meaning that is crucial.

These findings support the social interactionist conception of technology. This in turn has important implications for the process of implementation as it

155

suggests that the introduction of a technological innovation such as a GIS will entail more than simply securing the technical operation of the equipment. It is inevitable that perspectives that focus exclusively on equipment will conceptualize implementation as a reasonably straightforward process, dependent largely on the technological worth of the innovation and the technical competence of the staff responsible for implementation. However, if technology is conceptualized in broader social and political terms, implementation becomes an organizational rather than purely technical process.

The most significant aspect of the introduction of a new technology into an organization is the extent to which it implies a change to existing practices, particularly the underlying approach to decision-making. The implications of this change are barely considered by the technological determinist perspective and, while acknowledged by the managerial rationalists, are assumed to be readily controlled, given the careful preparation of logical strategies. However, as Handy (1993) has noted, organizations respond psychologically, not logically, to the process of implementing any form of change. Consequently, the identification of sources of power and methods of influence are likely to be of far greater value in the quest for effective utilization than rational argument.

Implicit within the technological determinist and managerial rationalist approaches to GIS implementation are assumptions about the role of spatial data in operational, managerial and strategic decision-making and therefore the needs of users. The instrumental rationality that underlies these perspectives leads to a presumption that the ability to integrate data from a variety of sources will facilitate data sharing and thereby result in better decisions and greater efficiency. Assumptions about the latent demand for data sharing both within as well as between organizations leads in turn to the corporate approach being regarded as the most appropriate framework for implementation.

There is a very close relationship between assumptions underlying the manner in which implementation should be approached and the role that information is expected to perform. In each case the technological determinist and managerial rationalist perspectives place emphasis on an idealized view of how organizations and decision-making ought to operate. The findings of the case studies, however, suggest that there are considerable dangers if those responsible for implementation do not start from an understanding of the particular values and practices of the organization in which the system will have to operate. The technology may be new but the organizational traditions and systems will have been refined and developed over many years and perhaps even generations. It is important to recognize, therefore, that to propose the corporate implementation of a GIS may imply a major structural change to the way that activities are conducted within many organizations. Moreover, a structural change of this kind is impossible without cultural change. Such were the circumstances in the majority of the case studies implementing systems involving several departments. In most instances there was little tradition of corporate working and it is not therefore surprising that deep-rooted resistance to the structural changes associated with the process of implementation was noted in some cases. It was evident that individuals within organizations were willing to accept the adoption of the technology but not necessarily the accompanying assumptions where these did not accord with the existing organizational culture.

These findings endorse the underlying sentiments of the social interactionist perspective, indicating the importance of taking an understanding of the nature of individual organizations as the starting point for implementation. Strategies that ignore the values and practices of organizations appear to court disaster. It therefore follows that there is not a single approach to implementation which guarantees effective utilization in all circumstances. It was evident in the case study organizations that different cultures will reach the same ends but by very different means (see Lynn 1990). The critical features of these approaches will be examined in the next section as such findings have significant implications for practice. It also follows from this finding that the outcome and impact of implementation will vary between organizations despite the acquisition of the same configuration of equipment (see Scott 1990). Given the earlier discussion, it is quite possible for the same machinery to have been purchased, yet perceptions of its potential value and role may differ markedly.

The analysis of the findings has so far concentrated on the cultural diversity between organizations. Such features can also be found within organizations, as there is often a wide variety of ideas and accepted practices. Implementation therefore has to be negotiated among a large number of interested parties, each of which will be assessing the extent to which the new technology affects their existing status and position. In circumstances where there are strong professional groupings, as in the case studies, there is likely to be a constant tension between the centre and the operational subsections. It is inevitable that a new technology will become embedded within these relationships and subject to their vagaries.

Overall the findings of this research tend to accord most closely with the explanatory framework provided by the social interactionist perspective. GIS implementation appears to be essentially a social and political process rather than simply a technological matter. Technical problems tend to reinforce existing organizational difficulties and are not in themselves responsible for the failure of the process of implementation. The association with organizational change links such a process to the underlying values and practices of the particular culture acquiring the system. The link between organizational culture and implementation emphasizes that change cannot be effectively introduced if it is imposed or controlled. GIS implementation, as in other areas, involves the much more subtle process of nurturing and cajoling change. The implications of these findings for practice will be explored in the next section.

Implications for practice

The extent to which the research findings support the assumptions underlying the social interactionist perspective challenges the appropriateness of the prescriptions that have dominated the GIS field. There are no quick fixes or recipes for success. In some senses the cookbook analogy is quite a good one for, as with cooking, despite following the recipe to the letter, the result often fails to live up to expectations. Soufflés fail to rise, sponges sink and sauces separate. The conceptualization of GIS implementation as a social and political process is crucial. This shifts the focus of attention from the technology to the organization. Furthermore, it emphasizes that for effective utilization to be achieved it is vital that the process of implementation starts from an understanding of how particular organizations operate in practice and not an idealized notion of how they should. Subsequent comments evaluate the experiences of the case studies, focusing in particular on the processes that appear to facilitate implementation.

The issue that dominated the discussion of implementation was how the case studies coped with change induced both by the technology itself as well as from a variety of other sources. Organizations that were able to deal with change were far more likely to secure utilization than environments where individuals had a tendency to take fright once problems started to be encountered. The manner in which the uncertainties associated with change and instability were dealt with appeared to be closely linked to the culture of the organization. However, it was possible to discern, from those that had achieved utilization or appeared likely so to do, three differing approaches. Each of these approaches reflected the underlying values and practices of the organizations concerned.

The first of these is summarized by the phrase 'innately innovative organizations'. Change, such as the introduction of new technology, appeared to be regarded within such organizations as an entirely natural part of their operations. This is not to suggest that there was an absence of debate, perhaps even quite violent debate, or that no problems were encountered, but rather than such issues threatening the whole future of the project they were viewed as an inherent part of the process of implementation. Out of the 12 case studies investigated, 2 appeared to have the capacity to take on the organizational changes implied by the introduction of GIS technology and sustain the process within a highly dynamic context. A critical feature of these organizations were the qualities of the staff. In both cases the overall skill levels and expertise were far higher than in other similar environments. These qualities tended to be combined with an inherent confidence in an overall strategic vision which was perceived to have benefits for both the individual as well as the organization. The personalities and skills of these individuals were undoubtedly important but it was the existing culture that appeared to be responsible for gathering these individuals into the organization. As a result, innovative environments would seem to attract innovative individuals and presumably the reverse is also the case. The close relationship between organizational culture and the staff appointed contributes significantly to the perpetuation of the existing values and practices of a particular organization. It follows on from this analysis that the two case studies that exemplified these features demonstrated a long tradition of innovation in every aspect of their activities including a wide range of policy fields as well as information management. The technological characteristics of the GIS projects in these two organizations were very different. However, more importantly they each exhibited a fundamental capacity to treat change as an opportunity. In both cases mistakes had been made but, rather than thwarting initiative, such errors tended to be regarded as an inevitable part of the process of organizational and personal learning, a process that is an essential element of implementation. Moreover, such values would seem to provide these environments with the flexibility necessary to sustain the implementation process and thereby ensure the long-term utilization of technological innovations like GIS.

The second approach that resulted in the utilization of the technology purchased stemmed from the opposite philosophy to the one cited above. In this case, while new ideas such as GIS were accepted, the aim of the process of implementation was to keep the disruption and need for change to an absolute minimum. The culture of such an organization was therefore marked by a profound sense of stability. In this case inertia should not be confused with stability. The organization that exemplified these characteristics among the case studies actually showed a willingness to embrace new ideas but without any great flourish. The fundamentals were identified and the instability necessary kept to a minimum. Moreover, it follows on from a capacity to create stability that such organizations also demonstrate the ability to sustain the process of GIS implementation. The results of this process tend not to be all that exciting in intellectual terms but the approach is indicative of a profound understanding of the nature of organizational cultures and the extent to which change has to be nurtured rather than imposed. Perceptive qualities of this depth appeared to be all too rare in the case studies.

The third, although far less certain, set of circumstances that appeared to contribute at least to short-term GIS utilization were the efforts of a single individual, most commonly referred to as a champion. Evidence from the case studies suggest that an enterprising and politically skilled individual can secure utilization, at least for a small project and a relatively short period of time. However, the critical issue is whether the organization can sustain the development and utilization of the innovation. The experiences of the case studies indicate that this is doubtful as the most expert individual cannot ensure effective implementation in a vacuum. Moreover, there is every chance that if the individual is out of step with the organizational culture they will eventually become so frustrated that they leave. This emphasizes the fragility of placing too much stress on the capabilities of one individual.

This evaluation points to the significant role of organizational cultures in the quest for effective GIS implementation. It is not necessarily the culture itself that determines the outcome but rather an understanding of the values and practices of the organization. The analysis of the experiences of the case studies also enables insight to be gained into how implementation was accomplished within these environments at a more detailed level. The initial overview of the social interactionist perspective suggested that three important components should be incorporated into any approach to implementation which starts by assuming the process to be social and political in nature. These are as follows:

1. an information management strategy that identifies the need of users and takes account of the resources and values of the organization;
2. commitment to and participation in the implementation of the system by individuals at all levels of the organization; and
3. an ability to cope with change.

Overall the findings of the research provide support for the underlying assumptions implicit within these recommendations for implementation. However, the detailed case study analysis suggests the need for some refinement. Those organizations that secured effective utilization of GIS technologies appeared to focus on the following issues in formulating their respective approaches to implementation:

1. the identification of simple applications producing information that was fundamental to the work of potential users;
2. an awareness for limitations of the organization in terms of accepted practices and available resources;
3. user-directed implementation set within a framework based on the commitment and real participation of staff throughout the organization; and
4. an ability to create stability or cope with change.

These four elements refine and develop the earlier framework, emphasizing the critical importance of change and the need to have an appreciation and understanding of the cultural values of the organization. In particular they demonstrate the importance of devising an information management strategy which identifies the core needs of users and the type of service that they require and link this to the accepted practices, traditions and resources of the organization. Consequently, information is viewed in terms of its social and political dimensions, not simply as a value-neutral object. The research findings suggest that the form of the strategy is largely irrelevant, that is to say whether it is a formal document or not. It is the process itself that is of far greater significance. Commitment and participation are also vital components of implementation. The experiences of the organizations studied suggest that token gestures at involving staff are likely to prove counter-productive. Moreover, it is crucial that implementation is set within a user-centred framework. The failure of technically operational systems to be integrated into the work of users in several of the case studies demonstrates the dangers of not fully involving these individuals within the process. The final component stresses the issue that is perhaps most fundamental, that is understanding the capacity of a particular organization to deal with change.

The effective implementation of a technological innovation such as a GIS is therefore dependent upon understanding the particular environment in which it will be expected to operate. It is crucial to start with an appreciation of the values and practices of the organization that are reflected in the needs of users, rather than taking some idealized notion of how organizations ought to operate. It is, however, rational conceptions of organizations that have dominated the GIS literature, resulting in the focus being placed on the technology and not the organizational context. The problems encountered by many of the case studies appear largely to result from accepting such approaches and thereby attempting to implement systems that appear alien to the culture in which they have been placed. If the changes necessary to implement the system fail to take account of the existing culture, there is a significant chance that the innovation will be rejected by users. It is difficult to avoid the agricultural analogy that if a crop is not planted and nurtured appropriately it is unlikely to survive. The same would seem to apply to the implementation of GIS.

Implications for research

The theoretical and practical implications of this research suggest the urgent need for greater understanding of the social and political dimensions of technological innovation. Furthermore, the findings indicate that much of the debate that has dominated the GIS field over the best management strategy to adopt for implementation is largely misguided. Effective implementation does not appear to be dependent upon the formulation of optimal strategies but rather tailoring the approach to the organizational circumstances in which the system will operate. Nor is this a once-and-for-all task. It is an ongoing process which at times may seem to be making little discernible progress. It is a fundamental aspect of implementation in that it represents the means by which change is introduced into a particular environment while at the same time being subject itself to changing circumstances.

In some respects these findings are by no means new. The problematic nature of the process of implementation has been demonstrated in studies focusing on computer technology as well as work investigating the failure of many policy initiatives to deliver the expected results in a wide range of national contexts. (See, for example, Barrett and Fudge 1981; Calhoun, Drummond and Whittington 1987; Campbell 1992b; Danziger *et al.* 1982; Danziger and Kraemer 1986; Eason 1988; Frissen 1989; Leonard-Barton 1988; Markus and Robey 1988). All these studies indicate the importance of conducting research that crosses traditional disciplinary fields. In the case of GIS it would appear that the chances of deriving practical benefits from this technology are likely to be limited if the focus of the research rests purely in computer science. Many of the problems in practice seem to be a direct consequence of ignoring the sociological, psychological, political and even anthropological aspects of such developments. In more specific terms there may be considerable theoretical insight to be gained from exploring the experiences of implementation in a variety of areas but perhaps most specifically policy implementation. It is also important in relation to GIS that further detailed studies are conducted in a range of organizational and national settings as well as investigations that seek to explore the impact of time on implementation. The findings of this work will enhance understanding of the processes affecting the development of particular technologies over time. It is possible that, while studies conducted in other fields tend to confirm the analysis presented here, the findings may simply reflect the circumstances pertaining at one point in time in a few organizations.

One of the most important aspects of these findings is the extent to which the culture of an organization seems to have a significant impact on its capacity to absorb change. It has only been possible for this study to scratch the surface of this issue. As a result there is a need for much greater understanding of the nature of organizations if resources are not to be wasted. It is, however, important that such studies do not solely concentrate on what might be termed the 'glamorous' environments, that is, those that this research has referred to as innately innovative. These are always likely to be in the minority. In many ways the more remarkable are those that have recognized their limitations and have had the perceptive capacities to gain from new developments through reducing the level of change necessary to a minimum. An important issue therefore is how organizations recreate themselves and to what extent it is possible to intervene in this process.

The findings of the research also question many assumptions about the role of geographic information in decision-making, which have tended to be taken for granted within the GIS field. In particular there needs to be much greater understanding of the role and nature of spatial data sharing within organizations. Many of the case studies indicated that, while they had adopted GIS technology, they had not also taken on board the underlying philosophy that is often associated with such systems. It is crucial therefore that those concerned with the development of GIS start with an understanding of how decisions are made in practice rather than a rational model that bears little relationship to the circumstances found in the real world.

The substantive fields that have been identified raise a further issue for the conduct of research in this area, namely the need for a greater variety of methodological approaches to be adopted. It will be impossible to gain enhanced understanding of the relationship between innovations and organizations without undertaking detailed case studies. Longitudinal research techniques also have a significant role to play in such work if real insight is to be gained into the way that organizations deal with change. This again points to the need to draw on approaches from a wide variety of disciplinary fields. Failure to embrace a full range of methodological techniques and theoretical insights does considerable disservice to the organizations that have to cope with the implementation of these technologies.

Conclusion

The relationship between organizations and technological innovations such as GIS is crucial to the outcome of the process of implementation. The decision as to whether GIS technology is or will be an innovation or an irrelevance will be decided by individual users working in a vast variety of organizational contexts. This study indicates that it is people that create and recreate their own worlds, not technology. Adoption of an innovation does not necessarily imply that those within the organization share the inventor's assumptions about how it should be applied. It is quite possible that some will find that it fails to accord with the values and practices of the organization and abandon further development. If those within the GIS field ignore the cultures into which their much-vaunted technology is expected to operate, it is unlikely that such systems will survive to see the next millennium. Innovations, like fashions, come and go but organizations will always exist and people will always be people.

References

ARGYRIS, C. (1971) Management information systems: the challenge to rationality and emotionality, *Management Sciences*, **17**, B275–B292.

ARGYRIS, C. AND SCHON, D. (1978) *Organisational Learning: A Theory of Action Perspective*, Reading: Addison-Wesley.

ASSIMAKOPOULOS, D. (1995) Greece: the development of a GIS community, in Masser, I., Campbell, H., Craglia, M. and Wegener, M. (Eds) *GIS Diffusion in Local Government in Europe*, London: Taylor & Francis.

AUDIT COMMISSION FOR LOCAL AUTHORITIES IN ENGLAND AND WALES (1990) *Management Papers: Preparing an Information Technology Strategy: Making IT Happen*, London: HMSO.

AUDIT COMMISSION FOR LOCAL AUTHORITIES AND THE NATIONAL HEALTH SERVICE IN ENGLAND AND WALES (1994) *High Risk/High Potential: A Management Handbook on Information Technology in Local Government*, London: HMSO.

BAINS COMMITTEE REPORT (1972) *The New Local Authorities: Management and Structure*, London: HMSO.

BARR, R. (1991) A federal approach to GIS, *Mapping Awareness*, **5** (6), 15–19.

BARRETT, S. AND FUDGE, C. (1981) *Policy and Action: Essays on the Implementation of Public Policy*, London: Methuen.

BARRETT, S. AND MCMAHON, L. (1990) Public management in uncertainty: a micro-political perspective of the health service in the United Kingdom, *Policy and Politics*, **18**, 257–68.

BATES, B.J. (1988) Information as an economic good: sources of individual and social value, in Mosko, V. and Wasko, J. (Eds) *The Political Economy of Information*, pp. 76–94, Madison: University of Wisconsin Press.

BATLEY, R. AND STOKER, G. (Eds) (1991) *Local Government in Europe: Trends and Developments*, London: Macmillan.

BEATH, C.M. (1991) Supporting the information technology champion, *MIS Quarterly*, **15**, 355–72.

BIJKER, W.E., HUGHES, T.P. AND PINCH, T.J. (Eds) (1987) *The Social Construction of*

Technological Systems: New Directions in the Sociology and History of Technology, Cambridge, MA: MIT Press.

BOLAND, R.S. AND HIRSCHHEIM, R.A. (Eds) (1987) *Critical Issues in Information Systems Research*, Chichester: John Wiley.

BRAVERMAN, H. (1974) *Labour and Monopoly Capital*, New York: Monthly Review.

BROMLEY, R. AND COULSON, M. (1989) The value of corporate GIS to local authorities: evidence of a needs study in Swansea City Council, *Mapping Awareness*, 3 (5), 32–5.

BROMLEY, R. AND SELMAN, J. (1992) Assessing readiness for GIS, *Mapping Awareness*, 6 (8), 9–12.

BRYMAN, A. (Ed.) (1988) *Doing Research in Organisations*, London: Routledge.

BUCHANAN, D.A. (1994) The organisational politics of technological change, in Medyckyj-Scott, D. and Hearnshaw, H.M. (Eds) *Human Factors in Geographical Information Systems*, pp. 211–22, London: Belhaven.

BUDIC, Z.D. (1994) Effectiveness of geographic information systems in local planning, *American Planning Association Journal*, 60, 244–63.

BURROWS, R. AND LOADER, B. (Eds) (1994) *Towards a Post-Fordist Welfare State?*, London: Routledge.

CALHOUN, C., DRUMMOND, W. AND WHITTINGTON, D. (1987) Computerised information management in a system-poor environment: lessons from the design and implementation of a computer system for the Sudanese Planning Ministry, *Third World Planning Review*, 9, 361–79.

CAMPBELL, H.J. (1990a) The use of geographic information in local authority planning departments, unpublished PhD thesis, University of Sheffield.

CAMPBELL, H.J. (1990b) The organisational implications of geographic information systems in British local government, in Harts, J., Ottens, H. and Scholten, H. (Eds) *Proceedings of the First European Conference on Geographical Information Systems*, pp. 145–57, Utrecht: EGIS Foundation.

CAMPBELL, H.J. (1991) Organisational issues in managing geographic information, in Masser, I. and Blakemore, M. (Eds) *Handling Geographic Information*, pp. 259–82, London: Longman.

CAMPBELL, H.J. (1992a) The impact of geographic information systems on British local government, *Computers, Environment and Urban Systems*, 16, 531–41.

CAMPBELL, H.J. (1992b) Organisational issues in the implementation of GIS in Massachusetts and Vermont: some lessons for the UK, *Environment and Planning B*, 19, 85–95.

CAMPBELL, H.J. (1993) GIS implementation in British local government, in Masser, I. and Onsrud, H.J. (Eds) *Diffusion and Use of Geographic Information Technologies*, pp. 117–146, Dordrecht: Kluwer.

CAMPBELL, H.J. (1995) Organisational cultures and the diffusion of GIS technologies, in Masser, I., Campbell, H.J., Craglia, M. and Wegener, M. (Eds) *GIS Diffusion in Europe*, London: Taylor & Francis.

CAMPBELL, H.J. AND MASSER, I. (1992) GIS in local government: some findings from Great Britain, *International Journal of Geographical Information Systems*, 6, 529–46.

CAMPBELL, H.J., MASSER, I., POXON, J. AND SHARP, E. (1994) *Monitoring the Take-up of GIS in British Local Government*, Luton: Local Government Management Board.

CHILD, J. (1985) Managerial strategies, new technology and labour process, in Willmott, H. and Collinson, D. (Eds) *Job Redesign: Critical Perspectives on the Labour*

Process, pp. 105–41, Aldershot: Gower.
CONSTANT II, E.W. (1987) The social locus of technological practice: community, system or organization?, in Bijker, W.E., Hughes, T.P. and Pinch, T.J. (Eds) *The Social Construction of Technological Systems: New Directions in the Sociology and History of Technology*, pp. 223–42, Cambridge, MA: MIT Press.
COULSON, M. AND BROMLEY, R. (1990) The assessment of user needs for corporate GIS: the example of Swansea Council, in Harts, J., Ottens, H. and Scholten, H. (Eds) *Proceedings of the First European Conference on Geographical Information Systems*, pp. 209–17, Utrecht: EGIS Foundation.
CROSSWELL, P.L. (1991) Obstacles to GIS implementation and guidelines to increase the opportunities for success, *Urban and Regional Information Systems Association Journal*, **3**, 43–56.
CURRY, M. (1995) Data protection and intellectual property: information systems and the Americanisation of the new Europe, *Environment and Planning A* (forthcoming).
DANZIGER, J.N., DUTTON, W.H., KLING, R. AND KRAEMER, K.L. (1982) *Computers and Politics: High Technology in American Local Government*, New York: Columbia University Press.
DANZIGER, J.N. AND KRAEMER, K.L. (1986) *People and Computers: The Impacts of Computing on End Users in Organizations*, New York: Columbia University Press.
DEAL, T.E. AND KENNEDY, A.A. (1982) *Corporate Cultures: The Rites and Rituals of Corporate Life*, Reading: Addison-Wesley.
DEPARTMENT OF THE ENVIRONMENT (1987) *Handling Geographic Information: Report of the Committee of Inquiry Chaired by Lord Chorley*, London: HMSO.
DOWNS, A. (1967) A realistic look at the final payoffs from urban data systems, *Public Administration Review*, **27**, 204–10.
DUNLOP, C. AND KLING, R. (Eds) (1991a) *Computerization and Controversy: Value Conflicts and Social Choices*, San Diego: Academic Press.
DUNLOP, C. AND KLING, R. (1991b) Social controversies about computerization, in Dunlop, C. and Kling, R. (Eds) *Computerization and Controversy: Value Conflicts and Social Choices*, pp. 1–12, San Diego: Academic Press.
DUNLOP, C. AND KLING, R. (1991c) The dreams of technological utopianism, in Dunlop, C. and Kling. R. (Eds) *Computerization and Controversy: Value Conflicts and Social Choices*, pp. 14–30, San Diego: Academic Press.
DUTTON, W.H., ROGERS, E.M. AND SUK-HO JUN (1987) The diffusion and impacts of information technology in households, *Oxford Surveys in Information Technology*, **4**, 133–93.
EASON, K.D. (1988) *Information Technology and Organisational Change*, London: Taylor & Francis.
EASON, K.D. (1993) Gaining user and organizational acceptance for advance information systems, in Masser, I. and Onsrud, H.J. (Eds) *Diffusion and Use of Geographic Information Technologies*, pp. 27–44, Dordrecht: Kluwer.
EASON, K.D. (1994) Planning for change: introducing a geographical information system, in Medyckyj-Scott, D. and Hearnshaw, H.M. (Eds) *Human Factors in Geographical Information Systems*, pp. 199–210, London: Belhaven.
ELMORE, R. (1978) Organizational models of social program implementation, *Public Policy*, **26**, 185–228.
EVELAND, J.D., KLEPPER, C. AND ROGERS, E.M. (1977) The innovation process in public organizations: some elements of a preliminary model, unpublished report, University of Michigan.

References

FEIGENBAUM, E. AND McCORDUCK, P. (1991) excerpts from The fifth generation: artificial intelligence and Japan's computer challenge to the world, in Dunlop, C. and Kling, R. (Eds) *Computerization and Controversy: Value Conflicts and Social Choices*, pp. 31–54, San Diego: Academic Press.

FELDMAN, M.S. AND MARCH, J.G. (1981) Information in organizations as signal and symbol, *Administrative Science Quarterly*, **26**, 171–86.

FLYNN, N. (1990) *Public Sector Management*, Hemel Hempstead: Harvester Wheatsheaf.

FRISSEN, P.H.A. (1989) The cultural impact of informatization in public administration, *International Review of Administrative Sciences*, **55**, 569–86.

GAULT, I. AND PEUTHERER, D. (1989) Developing GIS for local government in the UK, presentation at the European Regional Science Association Congress, Cambridge.

GIDDENS, A. (1979) *Central Problems in Social Theory*, Berkeley: University of California Press.

GIULIANO, V. (1991) The mechanisation of office work, in Dunlop, C. and Kling, R. (Eds) *Computerization and Controversy: Value Conflicts and Social Choices*, pp. 200–12, San Diego: Academic Press.

GOODMAN, P. (1993) Implementation of new information technology, in Masser, I. and Onsrud, H.J. (Eds) *Diffusion and Use of Geographic Information Technologies*, pp. 35–57, Dordrecht: Kluwer.

GOODMAN, P.S. AND SPROULL, L.S. AND ASSOCIATES (Eds) (1990) *Technology and Organizations*, San Francisco: Jossey-Bass.

HÄGERSTRAND, T. (1952) The propagation of innovation waves, *Lund Studies in Geography*, **4**.

HÄGERSTRAND, T. (1967) *Innovation Diffusion as a Spatial Process*, Chicago: University of Chicago Press.

HAMBLETON, R. AND HOGGETT, P. (1987) Beyond bureaucratic paternalism, in Hoggett, P. and Hambleton, R. (Eds) *Decentralisation and Democracy: Localising Public Services*, SAUS Occasional Paper 28, pp. 9–28, Bristol: SAUS, University of Bristol.

HANDY, C.B. (1991) *Gods of Management*, London: Souvenir.

HANDY, C.B. (1993) *Understanding Organisations*, Harmondsworth: Penguin.

HARRISON, R. (1972) Understanding your organization's character, *Harvard Business Review*, **50**, 119–28.

HAYNES, R.J. (1980) *Organisation Theory and Local Government*, London: George Allen & Unwin.

HIRSCHHEIM, R.A. (1985) *Office Automation: A Social and Organisational Perspective*, Chichester: John Wiley.

HIRSCHHEIM, R., KLEIN, H. AND NEWMAN, M. (1987) Information system development as social action: theory and practice, *RDP 87/6*, Oxford: Oxford Institute of Information Management, University of Oxford.

HOFSTEDE, G. (1980) *Culture's Consequences*, Beverly Hills: Sage.

HOGGETT, P. (1987) A farewell to mass production? Decentralisation as an emergent private and public sector paradigm, in Hoggett, P. and Hambleton, R. (Eds) *Decentralisation and Democracy: Localising Public Services*, SAUS Occasional Paper 28, pp. 215–33, Bristol: SAUS, University of Bristol.

HOGGETT, P. (1991) A new management in the public sector?, *Policy and Politics*, **19**, 243–56.

HOWARD, R. (1985) *Brave New Workplace*, New York: Viking Penguin.

INNES, J.E. AND SIMPSON, D.W. (1993) Implementing GIS for planning: lessons from the history of technological innovation, *American Planning Association Journal*, **59**, 230–6.

JAMES, P. AND POPE, R. (1993) Corporate GIS in local government: the Horsham experience, *Mapping Awareness*, **7** (4), 28–30.

KANTER, R.M. (1983) *The Change Masters: Corporate Entrepreneurs at Work*, London: Routledge.

KANTER, R.M. (1990) *When Giants Learn to Dance: Mastering the Challenge of Strategy, Management and Careers in the 1990s*, London: Unwin Hyman.

KEARNEY, A.T. (1990) *Barriers to the Successful Application of Information Technology: A Management Perspective*, Department of Trade and Industry and Chartered Institute of Management Accountants, London: HMSO.

KEEN, P.G.W. (1981) Information systems and organisational change, *Communications of the Association for Computing Machinery*, **24**, 24–33.

KING, J.L. (1985) Local government use of information technology: the next decade, in Toegras, C. (Ed.) *Managing New Technologies: The Information Revolution in Local Government*, pp. 11–35, Washington DC: ICMA.

KLEIN, H.K. AND HIRSCHHEIM, R.A. (1989) Legitimation in information systems development: a social change perspective, *Office, Technology and People*, **5**, 29–46.

KLING, R. (1980) Social analysis of computing: theoretical perspectives in recent empirical research, *Computing Surveys*, **12**, 61–110.

KLING, R. (1987) Defining the boundaries of computing across complex organisations, in Boland, R.J. and Hirschheim, R.A. (Eds) *Critical Issues in Information Systems Research*, pp. 307–62, Chichester: John Wiley.

KLING, R. (1991) Computerization and social transformations, *Science, Technology and Human Values*, **16**, 342–67.

KLING, R. AND IACONO, S. (1984) The control of information systems development after implementation, *Communications of the Association for Computing Machinery*, **27**, 1218–26.

KLING, R. AND SCACCHI, W. (1982) The web of computing: computer technology as social organization, *Advances in Computers*, **21**, 1–90.

KRAEMER, K.L. (1991) Strategic computing and administrative reform, in Dunlop, C. and Kling, R. (Eds) *Computerization and Controversy: Value Conflicts and Social Choices*, pp. 167–80, San Diego: Academic Press.

KRAEMER, K.L. AND KING, J.L. (1986) Computing and public organisation, *Public Administration Review*, **46**, 488–96.

LADD, J. (1991) Computers and more responsibility: a framework for an ethical analysis, in Dunlop, C. and Kling, R. (Eds) *Computerization and Controversy: Value Conflicts and Social Choices*, pp. 664–75, San Diego: Academic Press.

LAKE, R.W. (1993) Planning and applied geography: positivism, ethics and geographic information systems, *Progress in Human Geography*, **17**, 404–13.

LARSEN, J.K. (1985) Effect of time of information utilisation, *Knowledge: Creation, Diffusion, Utilisation*, **7**, 143–59.

LEACH, S., STEWART, J. AND WALSH, K. (1994) *The Changing Organisation and Management of Local Government*, Basingstoke: Macmillan.

LEONARD-BARTON, S. (1988) Implementation characteristics of organizational innovations: limits and opportunities for management strategies, *Communications Research*, **15**, 606–31.

LONG, R. (1987) *New Office Information Technology: Human and Managerial Implications*, London: Croom-Helm.
LOPEZ, X.R. AND JOHN, S. (1993) Data exchange and integration in UK local government, *Mapping Awareness*, **7** (5), 37–40.
LUCAS, H. (1975) *Why Information Systems Fail*, New York: Columbia University Press.
LYNN, L.H. (1990) Technology and organizations: a cross-national analysis, in Goodman, P.S., Sproull, L.S. and Associates (Eds) *Technology and Organizations*, pp. 174–99, San Francisco: Jossey-Bass.
LYYTINEN, K. AND HIRSCHHEIM, R. (1987) Information systems failures revisited: a survey and classification of the literature, *Oxford Surveys in Information Technology*, **4**, 257–309.
MAGUIRE, D.J., GOODCHILD, M.F. AND RHIND, D.W. (Eds) (1991) *Geographical Information Systems: Principles and Applications*, 2 vols, London: Longman.
MAHONEY, R.P. (1989) Should local authorities use a corporate or departmental GIS? *Mapping Awareness*, **3** (2), 57–9.
MAHONEY, R.P. AND MCLAREN, R.A. (1993) Best practice guidelines for GIS implementation in GB local government, *Proceedings of the Association for Geographic Information Conference*, pp. 1.20.1–1.20.6, Rickmansworth: Westrade Fairs Ltd.
MARCH, J.G. (1987) Ambiguity and accounting: the elusive link between information and decision making, *Accounting, Organisations and Society*, **12**, 153–68.
MARCH, J.G. AND OLSEN, J.P. (1983) Organizing political life: what administrative reorganization tells us about government, *The American Political Science Review*, **77**, 281–96.
MARCH, J.G. AND SPROULL, L.S. (1990) Technology, management and competitive advantage, in Goodman, P.S., Sproull, L.S. and Associates (Eds) *Technology and Organizations*, pp. 144–73, San Francisco: Jossey-Bass.
MARKUS, M.L. (1983) Power, politics and MIS implementation, *Communications of the Association for Computing Machinery*, **26**, 430–44.
MARKUS, M.L. (1984) *Systems in Organizations: Bugs and Features*, Boston: Pitman.
MARKUS, M.L. AND BJØRN-ANDERSEN, N. (1987) Power over users: its exercise by systems professionals, *Communications of the Association for Computing Machinery*, **30**, 498–504.
MARKUS, L. AND ROBEY, D. (1988) Information technology and organisational change: causal structure in theory and research, *Management Science*, **34**, 583–98.
MASSER, I. (1992) Organizational factors in implementing urban information systems, *Proceedings of the Urban Data Management Symposium*, pp. 17–25, Delft: UDMS.
MASSER, I. (1993) The diffusion of GIS on British local government, in Masser, I. and Onsrud, H.J. (Eds) *Diffusion and Use of Geographic Information Technologies*, pp. 99–115, Dordrecht: Kluwer.
MASSER, I. AND CAMPBELL, H. (1994) Information sharing and the implementation of geographic information systems: some key issues, in Worboys, M. (Ed.) *Innovations in GIS*, pp. 217–27, London: Taylor & Francis.
MASSER, I. AND CAMPBELL, H. (1995) Information sharing: the effect of GIS on British local government, in Onsrud, H. and Rushton, G. (Eds) *Institutions Sharing Geographic Information*, Piscataway, NJ: Rutgers University Press.
MASSER, I. AND ONSRUD, H.J. (Eds) (1993) *Diffusion and Use of Geographic Information Technologies*, Dordrecht: Kluwer.
MAUD REPORT (1967) *Local Government Administration in England and Wales*, London: HMSO.

McAusland, S. and Summerside, A. (1993) First steps to success, *Proceedings of the Association for Geographic Information Conference*, pp. 1.17.1–1.17.3, Rickmansworth: Westrade Fairs Ltd.
McRae, H. (1993) Taking the bull by the horns, *The Independent*, 12 March.
Medyckyj-Scott, D. and Hearnshaw, H.M. (Eds) (1994) *Human Factors in Geographical Information Systems*, London: Belhaven.
Meyer, J.W. and Rowan, B. (1977) Institutionalized organizations' formal structure as myth and ceremony, *American Journal of Sociology*, **83**, 340–63.
Miles, R. (1990) A stitch in time, *Computing*, 11 October, 22–3.
Moore, G.C. (1993) Implications from MIS research for the study of GIS diffusion: some initial evidence, in Masser, I. and Onsrud, H.J. (Eds) *Diffusion and Use of Geographic Information Technologies*, pp. 77–94, Dordrecht: Kluwer.
Moore, K. (1994) GIS implementation in local government: the Central approach, *Proceedings of the Association for Geographic Information Conference*, pp. 20.1.1–20.1.5, Rickmansworth: Westrade Fairs Ltd.
Morgan, G. (1986) *Images of Organization*, Beverly Hills: Sage.
Morgan, G. (1989) *Creative Organisation Theory*, London: Sage.
Mosco, V. and Wasko, J. (Eds) (1988) *The Political Economy of Information*, Madison: University of Wisconsin Press.
Mouritsen, J. and Bjørn-Andersen, N. (1991) Understanding third wave information systems, in Dunlop, C. and Kling, R. (Eds) *Computerization and Controversy: Value Conflicts and Social Choices*, pp. 308–20, San Diego: Academic Press.
Mowshowitz, A. (1976) *The Conquest of Will: Information Processing in Human Affairs*, Reading: Addison-Wesley.
Mumford, E., Hirschheim, R., Fitzgerald, G. and Wood-Harper, A.T. (Eds) (1985) *Research Methods in Information Systems*, Amsterdam: North-Holland.
Mumford, E. and Pettigrew, A. (1975) *Implementing Strategic Decisions*, London: Longman.
Naisbitt, J. (1984) *Megatrends*, London: McDonald.
Noble, D.F. (1984) *Forces of Production: A Social History of Industrial Automation*, New York: Knopf.
Norton, A. (1991) Western European local government in comparative perspective, in Batley, R. and Stoker, G. (Eds) *Local Government in Europe: Trends and Developments*, pp. 21–40, London: Macmillan.
Onsrud, H.J. and Pinto, J.K. (1991) Diffusion of geographic information innovations, *International Journal of Geographical Information Systems*, **5**, 447–67.
Onsrud, H.J. and Rushton, G. (Eds) (1995) *Institutions Sharing Geographic Information*, Piscataway, NJ: Rutgers University Press.
Openshaw, S., Cross, A., Charlton, M., Brunsdon, C. and Lillie, J. (1990) Lessons learnt from a post-mortem of a failed GIS, *Proceedings of the Association for Geographic Information Conference*, pp. 2.3.1–2.3.5, Rickmansworth: Westrade Fairs Ltd.
Openshaw, S. and Goddard, J.B. (1987) Some implications of the commodification of information and the emerging information economy for applied geographic analysis in the United Kingdom, *Environment and Planning A*, **19**, 1428–39.
Osborne, D. and Gaebler, T. (1992) *Reinventing Government: How the Entrepreneurial Spirit is Transforming the Public Sector*, Reading, Mass.: Addison-Wesley.
Paterson Report (1973) *The New Scottish Local Authorities: Organisation and Management Structures*, Edinburgh: HMSO.

PAYNE, D.W. (1993) GIS markets in Europe, *GIS Europe*, **2** (10), 20–2.
PETERS, T.J. AND WATERMAN, R.H. (1982) *In Search of Excellence: Lessons from America's Best-run Companies*, New York: Harper & Row.
PETTIGREW, A.M. (1985) *The Awakening Giant: Continuity and Change in ICI*, Oxford: Basil Blackwell.
PETTIGREW, A.M. (1988a) *The Management of Strategic Change*, Oxford: Basil Blackwell.
PETTIGREW, A.M. (1988b) Longitudinal field research on change: theory and practice, presentation at The National Science Foundation Conference on Longitudinal Research Methods, Austin, Texas, September.
PETTIGREW, A., FERLIES, E. AND MCKEE, L. (1992) *Shaping Strategic Change*, London: Sage.
PEUQUET, D.J. AND BACASTOW, T. (1991) Organizational issues in the development of geographical information systems: a case study of US Army topographic information automation, *International Journal of Geographic Information Systems*, **5**, 303–19.
PFEFFER, J. (1978) *Organizational Design*, Arlington Heights, Ill.: AHM Publishing.
PFEFFER, J. (1981) *Power in Organisations*, London: Pitman.
PFEFFER, J. (1992) *Managing with Power: Politics and Influence in Organizations*, Boston: Harvard Business School Press.
PICKLES, J. (1991) Geography, GIS and the surveillant society, *Proceedings of the Applied Geography Conference*, Toledo, Ohio.
PICKLES, J. (Ed.) (1995) *Ground Truth: The Social Implications of Geographical Information Systems*, New York: Guilford Press.
PINCH, T.J. AND BIJKER, W.E. (1987) The social construction of facts and artefacts: or how the sociology of science and the sociology of technology might benefit each other, in Bijker, W.E., Hughes, T.P. and Pinch, T.J. (Eds) *The Social Construction of Technological Systems: New Directions in the Sociology and History of Technology*, pp. 17–50, Cambridge, MA: MIT Press.
POSTMAN, N. (1992) *Technopoly: The Surrender of Culture to Technology*, New York: Knopf.
ROBEY, D. (1987) Implementation and organisational impacts of information systems, *Interfaces*, **17**, 72–84.
ROGERS, E.M. (1983) *Diffusion of Innovations*, New York: Free Press.
ROGERS, E.M. (1986) *Communication Technology: The New Media in Society*, New York: Free Press.
ROGERS, E.M. (1993) The diffusion of innovations model, in Masser, I. and Onsrud, H.J. (Eds) *Diffusion and Use of Geographic Information Technologies*, pp. 9–24, Dordrecht: Kluwer.
ROSZAK, T. (1994) *The Cult of Information: A Neo-Luddite Treatise of High-Tech, Artificial Intelligence, and the True Art of Thinking*, Berkeley: University of California Press.
RULE, J. AND ATTEWELL, P. (1991) What do computers do?, in Dunlop, C. and Kling, R. (Eds) *Computerization and Controversy: Value Conflicts and Social Choices*, pp. 131–49, San Diego: Academic Press.
RYAN, B. AND GROSS, N.C. (1943) The diffusion of hybrid seed corn in two Iowa communities, *Rural Sociology*, **8**, 15–24.
SCHEIN, E.H. (1980) *Organizational Psychology*, London: Prentice Hall.
SCHEIN, E.H. (1985) *Organizational Culture and Leadership: A Dynamic View*,

San Francisco: Jossey-Bass.
Scott, W.R. (1990) Technology and structure: an organizational-level perspective, in Goodman, P.S., Sproull, L.S. and Associates (Eds) *Technology and Organizations*, pp. 109–43, San Francisco: Jossey-Bass.
Simon, H.A. (1952) A behavioural model of rational choice, *Quarterly Journal of Economics*, **69**, 99–118.
Simon, H.A. (1973) Applying information technology to organisation design, *Public Administration Review*, **33**, 268–78.
South West Thames Regional Health Authority (1993) *Report of the Inquiry into the London Ambulance Service*, London: South West Thames Regional Health Authority.
Sproull, L.S. and Goodman, P.S. (1990) Technology and organizations: integration and opportunities, in Goodman, P.S., Sproull, L.S. and Associates (Eds) *Technology and Organizations*, pp. 254–65, San Francisco: Jossey-Bass.
Taylor, F.W. (1947) *Scientific Management*, New York: Norton.
Toffler, A. (1980) *The Third Wave*, New York: Random House.
van Buren, T.S. (1991) Rural town geographic information: issues in integration, *Proceedings of the Annual Meeting of URISA*, vol. 3, pp. 136–51, Washington: URISA.
Van de Ven, A.H. and Rogers, E.M. (1988) Innovations and organisations: critical perspectives, *Communications Research*, **15**, 632–51.
Weber, M. (1947) *The Theory of Social and Economic Organization*, trans. Hendersen, A.M. and Parsons, T., New York: Free Press.
Weick, K.E. (1990) Technology as equivoque: sensemaking in new technologies, in Goodman, P.S., Sproull, L.S. and Associates (Eds) *Technology and Organizations*, pp. 1–44, San Francisco: Jossey-Bass.
Weiss, C.H. (Ed.) (1977) *Using Social Science Research in Public Policy Making*, Lexington, MA: Lexington Books, D.C. Heath and Co.
Whisler, T.L. (1970) *The Impact of Computers on Organizations*, New York: Praeger.
Winter, P. (1991) Selling a corporate GIS, *Proceedings of the Association for Geographic Information Conference*, pp. 2.6.1–2.6.5, Rickmansworth: Westrade Fairs Ltd.
Worrall, L. (1994) Issues in the transfer of GIS into public sector policy making, in Bond, D., Reid, J., Stevens, M. and Worrall, L. (Eds) *GIS, Spatial Analysis and Public Policy*, Coleraine: University of Ulster, pp. 137–47.
Wright, S. (1994) *Anthropology of Organisations*, London: Routledge.
Yin, R.K. (1982) Studying phenomenon and context across sites, *Administration Science Quarterly*, **26**, 58–63.
Yin, R.K. (1994) *Case Study Research: Design and Methods*, 2nd Edn, Thousand Oaks: Sage.
Zuboff, S. (1988) *In the Age of the Smart Machine: The Future of Work and Power*, Oxford: Heinemann.

Index

accessibility of GIS 127
accountability of local authorities 58, 60
accuracy of information 125
Alper GIS 73, 78
applications of GIS 104–8
architecture department, GIS facilities 94
Arc/Info 73, 78, 79
aspect system, culture as 18
attribute data 103–4
automated mapping facilities 107
Axis 73, 78

background instability 144–5, 147–9
benefits, GIS 74–5, 78, 79
black-box quality of computers 13
building control department, GIS facilities 94
buildings department, GIS facilities 71
bureaucracy 57–8
    decision-making 20, 21
    style of 16
bureaucratic paternalism 57
business units 59

capitalism 19
case studies
    information management strategies 119–20
    merits 61
    research strategy 62–3
Census data 122
central government, and instability 145–6
centralization of local authority structure 57
champions 139–41, 143, 159
    charisma of 39
    social interactionism 39
change
    implementation of GIS 156, 158
    management of 154
    pressures for 144–5
    social interactionism 36, 41–2, 48
        agents of *see* champions
        management of 39

chief executive department, GIS facilities 70, 71
chief executives 57
    information technology, introduction 89
Chorley Commission 2, 4, 153
classically corporate system implementation 81–2
    case studies 85
cleansing department, GIS facilities 94
coalitions within organizations 17
commitment to GIS 132, 142–4
    facilitating mechanisms 139–42
    organizational context 133–9
    social interactionism 47–8
communication channels 4, 5, 6
competition
    local authorities 58
    within organizations 17
compulsory competitive tendering (CCT) 58–9, 87, 99–100, 123
    for GIS support 137
computer-aided design (CAD) 72
computer services department 70, 71, 94
conforming adolescent stereotype 101
Conservative governments, and bureaucracy 58
consistency in organizations 16
construction, data-processing approach 30
context 14
contingency factor, culture as 17–18
contracting out 95–6
cookbook method of system implementation 31
core–periphery model of diffusion 80
corporate approach
    GIS implementation 32, 33–4, 41–5, 81–2
        case studies 85
        organizational characteristics 99, 101
    local government, internal management 57
corporate information, data sharing 121
correspondence failure 49, 110–11

173

Index

cost-centre-based management structures 59, 123, 146
creative thinking 44
cultural phenomenon, organization as 18
culture, of organization 17–19
    decision-making style 20
currency of information 125
customization 28, 115–16, 130

data 104–5
    limitations 97, 116–17
data-processing approach 30
data-related problems, GIS 75, 78, 79
data sharing
    information management strategies 121–4
    managerial rationalism 32
    social interactionism 42–3
decentralization of local authorities 58–9
decision-making
    applications of GIS 104–6
    approaches to 16, 19–22
    GIS benefits 74, 75, 78, 79
    managerial rationalism 30
    social interactionism 44
departmental systems 69–72, 76, 81, 93–4, 99–100
Department of Transport 148
development department, GIS facilities 70, 71, 72, 77–8, 94
diffusion of innovations 1, 4–7
    GIS 65–7
        benefits and problems 74–6
        changes since 1991 76–9
        choice of technology 72–4
        evaluation 79–83
        system development 68–72
    implementation 7–8
    managerial rationalism 30
    social interactionism 36–7
    technological determinism 27
digital map data 103
digitizers 94
discrete-entity research strategies 61
district authorities *see* metropolitan districts
domination in organizations 20, 21–2

education 55
education department, GIS facilities 71, 94
efficiency, local authorities 56–7
    and cost 59
elections 148
electronics industry, effects of recession 147
emergency services 55
emergency services department, GIS facilities 72
enablement 59
energy-related services 55
engineers department, GIS facilities 70, 71, 72, 94

England, structure of local government 52, 55, 56
environmental health department, GIS facilities 71, 94
espoused theory 17
estates department, GIS facilities 70, 71, 72, 94
ethics, as element of technology 11
expectation failure 49, 111
experience with GIS 68, 69

failure
    of GIS implementation 110–12
    of innovations 11–12, 49–50
        causes 35
        technological determinism 28
feasibility studies 30
federal approach, GIS implementation 32–3
fiercely independent system implementation 82–3, 102
    case studies 85
fire services 55
functional
    clusters 122
    specialization 57

GDS 73
geographic information systems (GIS) 1–4, 153–4, 162–3
    diffusion 6, 65–7
        benefits and problems 74–6
        changes since 1991 76–9
        choice of technology 72–4
        evaluation 79–83
        system development 68–72
    implementation 113–14, 151–2, 154–5
        commitment and participation 132–44
        information management strategies 118–32
        instability 144–51
        managerial rationalism 31–4
        organizational considerations 118
        social interactionism 41–6
        technological considerations 114–18
    in practice 84–5
        case studies 85–6
        evaluation 108–12
        reasons for adoption 87–90
        technology characteristics 90–108
    reinvention 23–4
    as technology 12–15
GFIS 78, 79, 92
G-GP 73, 78
GIS *see* geographic information systems
Greater London Council 73

hardware
    as element of technology 10
    GIS 74, 78
        characteristics 91

Index

limitations 116
reliability problems 75
health care facilities 55
hierarchical model of diffusion 80
highways department, GIS facilities 70, 71, 72, 94
   instability 148
highways management system 97
horizontal integration 133–6
housing 55
housing department, GIS facilities 71, 94
human component of GIS 95–8
   staff turnover 148–9

imagery of technology 38, 88
implementation of technology 6, 7–8, 25–6, 154–5
   GIS 113–14, 151–2
   commitment and participation 132–44
   information management strategies 118–32
   instability 144–51
   organizational considerations 118
   technological considerations 114–18
   managerial rationalism 29–34
   social interactionism 34–48
   success and failure 49–50
   technological determinism 26–9
   typology 81–3
independence, professional 59
informal plans 40
information
   about GIS technologies 90
   requirements 124–6
information management strategies 118–19, 131–2
   case studies 119–20
   data sharing 121–4
   information requirements 124–6
   reasons for adoption of GIS 88
   resource implications of GIS 128–31
   service provided by GIS 126–7
   social interactionism 46–7
   technology 127–8
   user needs 121
information processing improvements 74, 75, 78, 79
information technology (IT)
   introduction 88–9
   strategies 119–20
information technology department, GIS facilities 70, 71, 94
instability 144
   background 147–9
   central government 145–6
   implications for GIS implementation 149–51
   pressures for change 144–5
   recession 147
interaction failure 49, 111

internal management, local authorities 56–60

knowledge
   as element of technology 10, 11
   social interactionism 37

land survey department, GIS facilities 94
land use planning 55
league tables 59
legal services department, GIS facilities 70, 71, 94
local government 51–2
   diffusion of GIS 65–7
      benefits and problems 74–6
      changes since 1991 76–9
      choice of technology 72–4
      evaluation 79–83
      system development 68–72
   internal management 56–60
   organizational aspects 22–3, 52
   research
      overview of findings 64
      strategy 61–3
   structure 52–6
London, structure of local government 52
London Stock Exchange, TAURUS system 6
loyalty in local government 57

machines
   as element of technology 10, 11
   organizations as 16
      decision-making style 20–1
mainframes 74, 78, 92
maintenance costs 96
management expertise 97–8, 130
managerialism 59
managerial rationalism 7, 29–31, 114
   case studies 87
   commitment and participation 132
   GIS adoption 79, 80, 81
   GIS applications 106
   implementation of GIS 156
   implications for GIS 31–4
   instability 144
   limitations 34–5
   organizational characteristics 101
MapInfo 73, 78, 79, 92
market forces 58
mass media 5
methods, as element of technology 10, 11
metropolitan districts 52–5
   approach to GIS implementation 69
   departmental GIS facilities 70, 71, 72
   GIS adoption 68
   GIS facilities 67
   hardware 74
   plans for GIS 66, 67
   software 73
microcomputers 74, 92

175

Index

minicomputers 74
momentum of GIS projects 136
multi-departmental systems 69–71, 72, 76, 81

National Health Service 59
neighbourhood offices 58
newly independent stereotype 102
non-metropolitan areas
    population 55
    structure of local government 55
Northern Ireland, structure of local government 52

office activities 27
on-the-job training 40
open-systems theory 17
    managerial rationalism 30
opinion leaders 5
Ordnance Survey data
    digital map 103
    problems 75, 116–17
    sharing 122
    Service Level Agreement 77, 79
organisms, organizations as 16
organizational characteristics of GIS 98–102
organizational problems, GIS 75, 78
organizations 51–2
    local authorities as 52
    nature of 13–23
    as unique cultures 38–9
OSCAR road network data 103–4
overspending 49

parks department, GIS facilities 70, 71, 94
participation in GIS 132, 142–4
    facilitating mechanisms 139–42
    organizational context 133–9
    social interactionism 47–8
passenger transport services 55
paternalism 101
    bureaucratic 57
peer opinion 5
people, as element of technology 10
performance indicators 59
person cultures 18
pilot projects 141–2, 143
planning department
    data 104
    GIS facilities 70, 71, 72, 77–8, 93–4
    organizational characteristics 98–9
    utilization 108
plotters 94
    quality 125
police services 55
policy, as element of technology 10
policy committees 57
political properties of information 43
politics within organizations 19
population of local authorities 55–6

postal surveys 62
post-bureaucratic management 60
power cultures 18
power within organizations 19
    decision-making 22
pragmatically corporate system implementation 82
    case studies 85
predictability in organizations 16
presentation of information 126
'pretty picture syndrome' 126
problems, GIS 75–6, 78, 79
process failure 49, 111
professional associations 14
    social interactionism 36
professional groupings
    counter-implementation strategies 42
    functional specialization 57–8
    horizontal integration 133
professional independence 59
pro-innovation bias 5
project selection 30
property department, GIS facilities 94
public health 55
public image of organizations 17

quality of hard copy 125

rebel adolescent stereotype 101–2
recession 147
recreation department, GIS facilities 70, 71
refuse collection 55
reinvention 23–4, 108–10, 155
requirements specification 30
research
    overview of findings 64
    strategy 61–3
resource implications of GIS 128–31
resources committees 57
return on investments 6
roads 55
role cultures 18, 19
rules, organizational 16

savings, GIS 74–5
Scandinavia 55–6
scientific management 16
    managerial rationalism 30
Scotland, structure of local government 52, 55, 56
Scottish districts 55
    approach to GIS implementation 69
    departmental GIS facilities 70, 71, 72
    GIS adoption 68
    GIS facilities 67
    hardware 74
    plans for GIS 66, 67
    population 55
    software 73

Index

Scottish Islands 55
  plans for GIS 66
  population 55
Scottish regions 55
  approach to GIS implementation 69
  departmental GIS facilities 70, 71, 72
  GIS adoption 68
  GIS facilities 67
  hardware 74
  plans for GIS 66, 67
  population 55, 56
  software 73
segmentalism 57
Service Level Agreement 77, 79
shire counties 55, 56
  approach to GIS implementation 69
  departmental GIS facilities 70, 71, 72
  GIS adoption 68
  GIS facilities 67
  hardware 74
  plans for GIS 66
  population 55, 56
  software 73
shire districts 55
  approach to GIS implementation 69
  departmental GIS facilities 70, 71, 72
  GIS adoption 68
  GIS facilities 67
  hardware 74
  plans for GIS 66, 67
  population 55, 56
  software 73
single departmental systems 69, 70, 72, 76, 81
social characteristics of technologies 38
social interactionism 7, 34–41, 114, 155
  case studies 87
  change and instability 48
  commitment and participation 47–8, 132
  GIS adoption 79, 80, 81
  GIS applications 106
  implementation of GIS 157
  implications for GIS 41–6
  information management strategy 46–7
  instability 144
socialization process 14
  local authorities 60
social services 55
social services department, GIS facilities 71
societal infrastructure, as element of technology 10
software 72–3, 78–9
  characteristics 91–2
  as element of technology 10
  limitations 115–16, 117
spatial data 44
spatial referencing 125–6
specialization, functional 57
staffing considerations, GIS 95–8
  turnover 148–9
standard instability *see* background instability

structure
  local government 52–6
  organizational 15, 19
subsystem, culture as 18
success
  as element of technology 11–12, 49–50
  in GIS implementation 110–12
surveyors department, GIS facilities 70, 94
  data 104
survival of the fittest 16
symbolic value of technology 38
  GIS 88
Sysdeco 73
system design 30
system development 68–72, 76
system implementation
  fiercely independent 82–3, 102
    case studies 85
  typology 81–3
    case studies 85
system modifications costs 96
systems analysis 30
systems, organizations as 17

task cultures 18–19
TAURUS system, London Stock Exchange 6
team building 40
technical characteristics of GIS 91–4
technical problems, GIS 75, 78
technical services department, GIS facilities 70, 71, 94
  data 104
techniques, as element of technology 10
technological determinism 7, 26–9, 114
  case studies 87
  GIS adoption 79, 80, 81
  GIS applications 106
  implementation of GIS 156
  limitations 34–5
technology 1–4, 9–12
  choice of 72–4
  implications for GIS 12–13
Teesdale 56
telephone-based survey approach 62
terminals with access to GIS 93
test sites 89, 92
theoretically corporate system
  implementation 82
    case studies 85
theory-in-use 17
training
  professional groupings 57
  resources 130
  social interactionism 40
transport services 55
trials, data-processing approach 30
Tyne and Wear 53–4

up-to-dateness of information 125
user-centred design philosophies 39–40, 41

177

Index

user needs 121
utilization of GIS 104–8

values, as element of technology 11
vendors 90
   instability 149
   recession, effects 147
vertical integration 136–9
visual display of information 126

Wales, structure of local government 52, 55, 56

water supply 55
web model of research 61, 63
West Germany 55–6
Wings 73, 78, 79
working groups 141, 143
works department, GIS facilities 71
workstations 74, 78, 92

X Assist 73